Selected Topics in Biology

Ken Hyde, Ph.D.

Harper College

Kendall Hunt
publishing company

Customized from *Zoology: An Inside View of Animals* by Ken Hyde.

Kendall Hunt
publishing company

www.kendallhunt.com
Send all inquiries to:
4050 Westmark Drive
Dubuque, IA 52004-1840

ISBN 978-0-7575-8838-9

Printed in the United States of America
10 9 8 7 6 5 4 3 2 1

Contents

Topic 1

The Enchanting Parade

*I*t seems to me that people are fixated on nature. Some are seduced by it, others are repelled. Think about it: rarely is someone lukewarm when it comes to things wild. For some it's an aversion or a phobia, such as a fear of spiders, snakes, centipedes, or tapeworms. They may prefer the sidewalk to the hiking trail, the motel to the campsite. For others there's a reverence. They are the bird-watchers, the whale enthusiasts, the butterfly chasers, the SCUBA divers, the fishermen. They may be active environmentalists, or may simply relish walking quietly along a wooded trail. Some may even become biologists.

So, in one way or another, people are involved with nature. Our innate being is somehow tied to the world around us. It's part of the human spirit. Some might go so far as to say it's woven into our genetic fabric. This fascination with nature even has a name: **biophilia**. Perhaps that's why you're in this course. You realize that you're a part of this enchanting parade and you want to learn more about how it came to be.

There is a tremendous, even awe-inspiring, diversity to the biological landscape. We will see that science explains this vast biodiversity by way of natural selection through differential reproduction. In other words, through evolution. So it seems that evolution then is the drummer to which organisms march in this incredible parade. But is that the only interpretation? Are there other ways to explain the appearance of so many different lifeforms? Certainly, more than one. In fact, each culture has its own version of the grand history of life. However, alternative viewpoints are, for the most part, religious in nature and as such lie well beyond the scope of science. These alternate explanations are, therefore, not detailed here. Not because they are unimportant, but rather because they belong to another domain, a very personal, yet separate, domain.

Yet, we cannot simply ignore alternate explanations altogether for, although separate, science and religion have a shared history—one that has, at times, been rather turbulent. As science advanced over the centuries, religion adjusted many of its interpretations to allow for scientific discoveries of the natural world. For instance, the role of miracles to explain unknown phenomena has given way as science uncovered nature's mysteries one after another. On the other hand, science has realized its limitations in explaining moral issues. Science, for instance, does not pretend to deal with such claims as a soul or life after death; those are purely the realm of religion.

Thus, each of us is left to form our own world view that somehow incorporates both science and religion. Over time, three such "world views" have emerged.

The Either-Or Group

Since it was first forwarded some 150 years ago, the theory of evolution has been the focus of considerable debate between the scientific and religious communities—a debate that is alive today in certain arenas. Many people, on both sides of the issue, claim that evolution and religion are incompatible, that they lead to conflicting conclusions. This is the either/or approach: either one or the other explanation is true, but not both. In this interpretation, since both explanations cannot be true, one must therefore deny the other.

This approach has led to the establishment of two opposing camps, each defending its own position while refuting the validity of the other; each entrenched behind its position as battle lines have been drawn, opposing positions taken, and volleys fired. In the United States, for example, creationists recently declared their explanation a "science" and, as such, demanded equal time in science curricula. Biologists, and the wider scientific community, responded with the view that creationism & Intelligent Design are religious positions, and, as such, cannot be viewed as science—a view the courts have upheld. And so, the two camps remain unyielding and far apart.

The Bridgebuilders

A contrary viewpoint is taken by many others who believe in a connecting link or bridge between evolution and religion. They propose that, in reality, there is confluence between the two; that science and religion are two elements of a Supreme Whole and each has a specific contribution to the understanding of that whole.

In this world view theologians claim that a Creator began the process on earth from which all life has sprung; however, it's up to science to explain the outcome. In other words, it's the evolutionists who will unravel the mechanism by which this creation is fulfilled. Religion then forms the framework for our understanding of life while science fills the gaps. Thus, one discipline supports the other. Here, one overall Truth prevails and both science and religion contribute their special insight to its resolution. According to the bridgebuilders, we can only understand life on earth if the two sides work together toward its true revelation. In the end, the One Truth will be upheld by two supporting buttresses: science *and* religion.

The Separatists

Then there are a number of scientists and theologians who have found satisfaction in the very idea that the two disciplines offer separate, but equal, lines of inquiry. In point of fact, say the separatists, both science and religion, seek answers to different, yet fundamentally profound questions. Science seeks to provide factual accounts of *how* we came to be, what is our history, and how we are related to other life-forms on our planet. Religion, on the other hand, addresses the question of *why* we came to be. It searches for the meaning behind our existence; it seeks to place values within the human context.

Both search, but in different realms. Science inspects the physical, tangible, factual universe while religion looks to the metaphysical, the philosophical, the ethical. Science relies upon experimentation, objective reality, direct observation, and reproducible tests to resolve its uncertainties while religion unveils its mysteries through revelation, personal faith, introspection, and moral discourse. These are both important endeavors, but different.

In the opinion of many scientists and theologians alike, science and religion have equally relevant messages. Both are, after all, seeking to answer universal questions, but with different tools, different approaches, different perspectives. Both are integral to the human experience, yet we should be careful not to confuse one with the other, they are not the same. One doesn't necessarily substantiate nor deny the other. If science is successful in its search to answer how we came to be, it doesn't, by default, answer the question of why we are here. Likewise, should religion reveal the why, it doesn't, by extension, answer the how question. As Stephen Gould[1] has so aptly put it, "the causes of life's history cannot solve the riddle of life's meaning."

We cannot deny the fact that both science and religion have had an immeasurable impact on human understanding. But they do so in different ways and for different reasons. However, they do come together in one respect... they both help us to appreciate the wonder of the world around us. So, regardless of the route you might follow in your discovery of the natural world, one thing is certain: animals and plants enrich the earth with their complexity, their matchless beauty, and their fascinating diversity. And in some way the human spirit, your spirit, is linked to that complexity, beauty, and diversity.

> ...REGARDLESS OF THE ROUTE YOU MIGHT FOLLOW IN YOUR DISCOVERY OF THE NATURAL WORLD, ONE THING IS CERTAIN: ANIMALS AND PLANTS ENRICH THE EARTH WITH THEIR COMPLEXITY, THEIR MATCHLESS BEAUTY, AND THEIR FASCINATING DIVERSITY.

[1]Gould, S. J. 1999. Rocks of Ages: Science and Religion in the Fullness of Life. New York, Ballantine Publishing.

Topic 2

Energy Dynamics

All living cells, without a single exception, require energy. No cell has much of a future without a steady stream of it; all organisms must be able to find it, collect it, consume it, and, finally, to assimilate it. Plants have solved the energy demand by transferring light energy into chemical energy, through photosynthesis. Animals, however, since they are made of cells and since they are *heterotrophic,* must rely upon an external source of food for their energy supply. The food, once **ingested**, undergoes either mechanical and/or chemical breakdown in a process called **digestion**. The products of digestion are then **assimilated** by individual cells. Once in the cell, the energy is then converted to **ATP** through the process of **cellular metabolism**.

The energy converting, metabolic process in the cell requires *oxygen* to complete its activity. In the presence of oxygen the energy is captured in the ATP molecule for later use. But during the process three by-products are liberated. First, some *water* is produced and the animal must finely regulate its concentration both inside and outside of the cells. Second, *carbon dioxide* is liberated; and finally, *nitrogen* is produced from the breakdown of amino acids present in digested proteins. Both carbon dioxide and nitrogen are toxic to the cells and must either be quickly removed or transferred into a nontoxic form. The diagram below represents the basic metabolic process within a typical cell.

Any undigested food is eliminated from the animal as **waste**. The names given to this waste material are many: feces, droppings, offal, excrement, stool, manure, sewage, as well as many not fit to mention here.

Feeding Mechanisms

As heterotrophs, animals have a number of feeding strategies available to them. If they feed primarily on plant material, they are known as **herbivores**. In contrast, animals that feed on other animals are called **carnivores**. However, while many animals prefer a herbivorous *or* a carnivorous diet, it is rare to find one that is exclusively one or the other; instead, most live as *omnivores*, feeding on both plant and animal materials. Animals that feed on dead flesh are **scavengers**, while those that prefer decaying material are **detritus feeders**.

Many invertebrate animals have two special feeding strategies at their disposal, **suspension** or **deposit** feeding. In suspension feeding, animals filter food from their aquatic surroundings. To do that requires water to be brought across a feeding surface (often lined by mucous as a trapping medium) followed by the removal of food particles such as bacteria, plankton, and detritus. Examples of suspension feeders include certain protozoa, sponges, some annelids, barnacles, and mollusks,

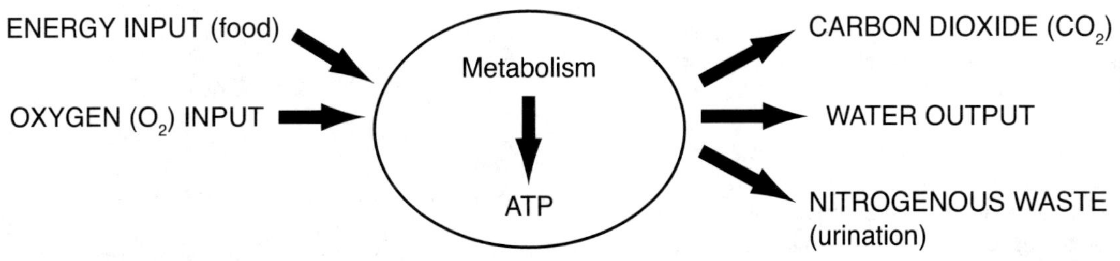

Figure 2.1 This summary diagram of cellular metabolism shows required material (food and oxygen) entering a typical cell from the left while by-products (water, CO_2, and nitrogenous wastes) exit from the right. The result of metabolism is the production of ATP molecules for use by the cell in its normal functions.

among others. Deposit feeders extract food from a soft substrate such as mud, soil, or sand. Those animals that simply swallow large portions of substrate are known as *direct deposit feeders* (e.g., earthworms and some snails); whereas, animals that carefully remove only the uppermost layer of a substrate are selective deposit feeders (e.g., sea cucumbers and many annelids).

A Light That Burns From Within

All work requires energy. In fact, the very definition of energy is the ability to do work. Whether it's the car in your garage, the toaster on your kitchen counter, the spider under your porch, or the cell buried deep within your *pectoralis major* muscle, energy is absolutely necessary if you want output. In fact, it's the most essential, indispensable ingredient in the machinery of life. And what do we know about energy?

(1) It cannot be created nor destroyed (at least that's what the physicists tell us); thus, it's not recyclable.

(2) The ultimate source of energy for living systems is the sun.

(3) Photosynthetic plants are needed to transform sunlight into chemical energy for their use and for ours.

(4) Cells, to stay alive, require a constant input of energy.

Oh, there's one more thing that we know about energy. Its use seems to be **inefficient**, in some cases, highly inefficient. Take, for example, the light bulb. Electric energy is needed to produce light from an incandescent bulb, but in so doing, around 98% of the energy input is lost as heat (just touch an angry light bulb and this fact is quickly evident).

Or, consider an entire ecosystem. As energy travels through one trophic level to another, heat is generated—about 90%, to be exact. That means that the producers (plants) and consumers (animals) use only about 10% of the energy available to them. The rest is carried away on the warm breath of Mother Nature.

Even cells are not all that efficient. The average liver cell, nerve cell, or muscle cell rarely exceeds 25% proficiency; the rest (~75% or so) is, again, lost as heat. You only need to cuddle with somebody or touch a person after a long, hard workout to appreciate the radiant factor of metabolism.

But not all living energy systems are so inefficient. There's one process that eclipses this seemingly universal trend of poor energy conversion. And it's a fascinating one indeed.

Bioluminescence it's called...the ability of living organisms to produce light. Its efficiency rating is *over 92%*—one of the most effective energy converters known! In fact, bioluminescence is often termed "cool light" reflecting its efficiency.

As we now understand it, bioluminescence is a chemical process in which dietary energy is transformed into radiant energy... light. To accomplish that, four things are needed: oxygen, ATP (the energy source), luciferin (a phosphorus-containing compound), and the enzyme luciferase. Thus, the formula for bioluminescence goes something like this:

oxygen
+
ATP
+
luciferin
+
luciferase
↓
LIGHT

The types of organisms using bioluminescence are many and they come from a wide variety of groups:

bacteria	**amoeba**
sponges	**fungi**
cnidarians	**bivalves**
cephalopods	**insects**
annelids	**fish**
crustaceans	**sharks**

A quick scan of the above list tells us that almost all bioluminescent organisms are aquatic and those that aren't are nocturnal in their terrestrial habits. To better understand the reason for that we need only to look at specific habitats.

First of all, approximately 90% of the earth is water. Of that, around 85% is below one mile, yet light only penetrates to about 600 feet. As a result, the vast majority of habitats on our planet are in total darkness, where no plants can exist.

The mesopelagic zone (below 600 feet, our largest environment) is a dark, uninviting, cold, pressurized, inhospitable place where the only food available to animals is each other or that which drifts down from above. Yet a wide variety of animals live there. And one of the factors that makes them able to do so is their ability to make light. Think about it. There aren't many ways to find food or each other in total darkness. But bioluminescence gives animals that ability.

There are two types of bioluminescence:

Extrinsic Bioluminescence

Many animals do not, or cannot, manufacture their own light and, thus, must "borrow" it from another organism. For instance, the flashlight fish has a pouch below each eye filled with bioluminescent bacteria. These symbiotic bacteria provide an illuminating glow that the fish uses to light up its surroundings.

Likewise, the angler fish has a dorsal spine that hangs in front of its large, viper like mouth. The spine has a terminal bulb filled with glowing symbiotic bacteria that serve as a lure to attract unsuspecting prey. There's even a squid that has illuminating bacteria in its ink sac. Imagine the surprise of a predator that becomes engulfed in a cloud of shimmering ink!

Intrinsic Bioluminescence

Most bioluminescent animals, however, can produce their own light through special light-producing organs called *photophores.* These special photogenic organs actually produce light through the same mechanism already described (luciferin + luciferase + ATP+ O_2= light).

This light producing organ (the photophore) has a small lens for focusing, a diaphragm-like iris that can change the size of the beam from a wide angle to a narrow pinpoint, and a mirror-like reflector that lines the back of the organ to magnify the light. In other words, the photophore is very much like an eye... *except that it sends light rather than receives it.*

We have discussed the nature of bioluminescence, how it's produced, and what kinds of animals use it. Now let's find out why they do so. Although there may be a number of specific reasons for an animal to use light, there are really only three basic uses: *to find food, to escape becoming someone else's food, and to communicate.*

In the first category, that of food finding or preying on another animal, an individual may simply illuminate its environment to locate a food source. The flashlight fishes do just that with two beacons of light emanating from their cheek pouches to spot food on the ocean floor. On the other hand, bioluminescence may be used *not* to illuminate your food but to attract it to you. Both the angler fish and the viper fish lure prey in this manner. They wiggle a tiny light in front of their gaping jaws and any curious fish that wants to "check out" the strange light will more than likely discover more than he's bargained for.

A large number of animals use bioluminescence in defense. They may startle a potential predator by a sudden flash or they may confuse, or even temporarily blind, a predator by issuing a cloud of bioluminescence ink as some squid are prone to do.

Another curious practice is seen in some animals that will autotomize (shed) a glowing body part when harassed. Some annelid (polynoid) worms, for instance, will purposely self-amputate a luminescent portion of their body when in the grasp of a predator. This draws attention to the flashing part while the rest of the worm escapes.

One more defensive strategy is a type of countershading called *ventral bioluminescence.* Located in the brain of some animals are special sensors termed **extra-ocular photoreceptors.** These receptors are designed to detect ambient light intensity, that is, the level of light in the animals' immediate surroundings. This information is then passed to ventral photophores, which mimic ambient light, thus, rendering the animal virtually invisible to potential prey lurking beneath.

Here's how this might work. Imagine a prowling shark looking overhead to find its prey by relying upon a dark outline silhouetted against a bright sky. The shark can then ambush the unsuspecting prey from below. However, if the prey is able to match its ventral surface to the surrounding light, its outline will become obscured and difficult, if not impossible, for the shark to see. Even greater camouflage is accomplished if the light flashes or glitters slightly to match the glimmer of sunlight on the water's surface.

Although it's hard to overestimate the importance of using light to escape enemies or to locate food, we, nonetheless, see even larger numbers of species using bioluminescence, not just for escape or food location, but to communicate. In fact, this may be its most valuable use.

Just recall summer evenings and the flashing of lightning bugs (fireflies) and you will realize that, even on land, bioluminescence has a role to play, and that role is communication.

To an animal that lives its life in darkness, either as a nocturnal terrestrial form or in the eternal blackness of the ocean floor, the ability to pass information quickly and unambiguously is critical. And light does just that.

The communicative functions of bioluminescence are many. Here are three of the more important ones.

(1) **Species Recognition.** Many animals have their own specific patterns of bioluminescence—lights unique to just them. As such they are easily recognized by other members of the same species. This can be important in establishing territories or in any type of group behavior.

(2) **Niche Distribution.** If emitting light enables an animal to establish a territory, this in turn, may help to distribute a species within available habitat and thus to prevent overcrowding and the overuse of local resources.

(3) **Reproduction.** But perhaps the most common use of bioluminescence is in the realm of reproduction. A specific pattern, color, or pulsation can identify the sex of the "flasher" while at the same time serve to attract a mate. Furthermore, light of this sort may be a crucial part of the courtship ritual.

Another reproductive use of light is in the glowing eggs of certain fish and polychaete worms which makes them easier for the males to find and fertilize.

Of course, animals do not always subscribe to an either-or agenda in their use of bioluminescence. They, instead, intertwine the various uses as the need arises. An individual animal may, for example, use bioluminescence to defend itself from a marauding predator one moment only to use it to hunt prey the next. Or, it may flash coded signals to induce group cohesion or to attract a mate.

In the darkness of night or the blackness of the marine abyss, light is employed in a variety of ways. To better understand its role in the life of animals, consider the following example:

The little flashlight fish, *Anomalops katoptron,* is a nocturnal species found in the coral reefs of the Pacific Ocean. It has a pouch beneath each eye in which bioluminescent bacteria are embedded. A flap that covers each pouch can be lowered to reveal glowing bacteria. In this way the fish uses the light as a beacon to locate food. However, it can also flash messages to other flashlight fish for reproductive purposes or just to help maintain proper distances from each other. In this way, *Anomalops* is using bioluminescence, not just for one function, but to achieve a combination of ends.

THE RESULT OF THIS SIMPLE CHEMICAL PROCESS CAN BE DAZZLING. LIKE A NEON BILLBOARD ITS CURIOUS FLASHES, THROBBING PULSES, AND BRILLIANT BURSTS OF ORANGE, YELLOW, AND BLUE LIGHT ARE EMITTED FROM A VARIETY OF ORGANISMS AND FOR A VARIETY OF REASONS.

Topic 3

Cellular Reproduction

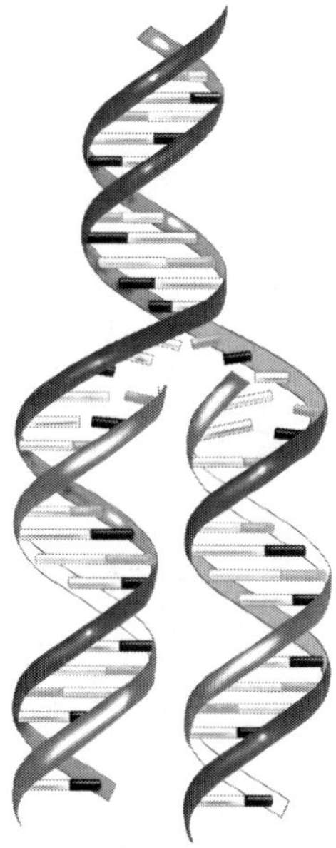

Figure 3.1 DNA Replication.

The Mitotic Marvel

What happens when you cut yourself? How does healing actually occur? How did you grow from a toddler to adulthood? How are the millions upon millions upon millions of cells that you lose each day replaced? What happens to your cells as you become older? Why can't you live forever? How does a primitive organism such as an amoeba reproduce? The answer to each of those questions lies in the fundamental biological process of mitosis, a process common to all known organisms. What is it about this process that makes it such a vital part of life?

Biologists universally agree that single-celled organisms were the ancestors to multicelled ones...to the rest of us. They have concluded, and rightfully so, that all higher animals came from simple single-celled protozoans. However, for a unicellular organism to become multicellular it must first find a way to make a copy of itself—an exact copy. If you don't quite understand why that's so, just imagine that you have burned your finger, resulting in the destruction of a number of epithelial cells. You want those dead cells replaced—but not by brain cells, not by liver cells, not even by muscle cells. You want them replaced by epithelial cells...you want your finger back the way it was. The events leading to the replacement of your finger cells lies at the very heart of mitosis.

We know that the instructions for life's many processes are wound in the genetic code. It's the DNA in our cells that's responsible for the structure and function of each cell, and by extension, for structure and function of the entire body. So, the answer to the problem of making perfect copies must be found in DNA. The first thing a cell must do to copy itself, therefore, is to make a copy of its DNA. That first step is called **replication**, chromosomal replication, to be exact. If a cell can make an unerring copy of its DNA, then the DNA will, in turn, direct each cell to carry out its specific functions. A skin cell behaves as a skin cell, a brain cell does what brain cells are supposed to do; DNA directs them along their proper paths. Not to belabor this point, but the key here is to make an exact copy of each and every chromosome. Replication is, in fact, the crucial event in the mitotic marvel.

Mitosis proceeds as a series of steps leading to the production of two new daughter cells genetically identical to each other. The first step, replication, occurs during cell interphase and results in the appearance of a double-stranded chromosome as diagrammed above. Each identical strand is a sister chromatid held together by a centromere. The actual process of mitosis, then, simply involves a sequence of events (prophase, metaphase, anaphase, and telophase) during which the sister chromatids are separated from each other and eventually passed on to two new, genetically identical cells (see diagram on next page).

Once the chromosomal material has been allocated to each daughter cell, the new cells can either separate to produce new organisms or they may remain together, continuing the process over and over, to produce a scar, or to replace damaged or lost cells, or to form a growing tissue.

The process is quite simple and reliable, as long as the chromosomes are replicated exactly. But mitosis is not without its difficulties. First of all, mutations can occur within the DNA molecule, which may lead to failed copies and potential disaster. Also, the process does not go on indefinitely. Quite the contrary. Have you noticed that as you age you take longer to heal? That cut on your finger that took a few days to heal when you were a youngster takes longer and longer to recover as you age. We think the reason for this is that DNA can only be copied so many times. Each replication slightly erodes the process until eventually cells lose their ability to reproduce altogether... a process termed **senescence**. Alas, we cannot live forever; the genetic code has seen to that! But our genes haven't abandoned us entirely. Instead, they have improvised a process that allows a part of us to live on. That process is *sexual reproduction through meiosis.*

The important thing to remember about mitosis is that, because of replication, each chromosome is duplicated **exactly**, which makes it possible for each new daughter cell to be genetically identical to each other. Since all multicellular organisms (including you and me) started life as a single cell (a zygote) and then relied upon mitosis to add cell after cell after cell, we can conclude that each cell in our body is genetically identical. That is true. In fact, it's the basis for cloning, for producing exact replicas of a living organism...a frog, a sheep, a human!

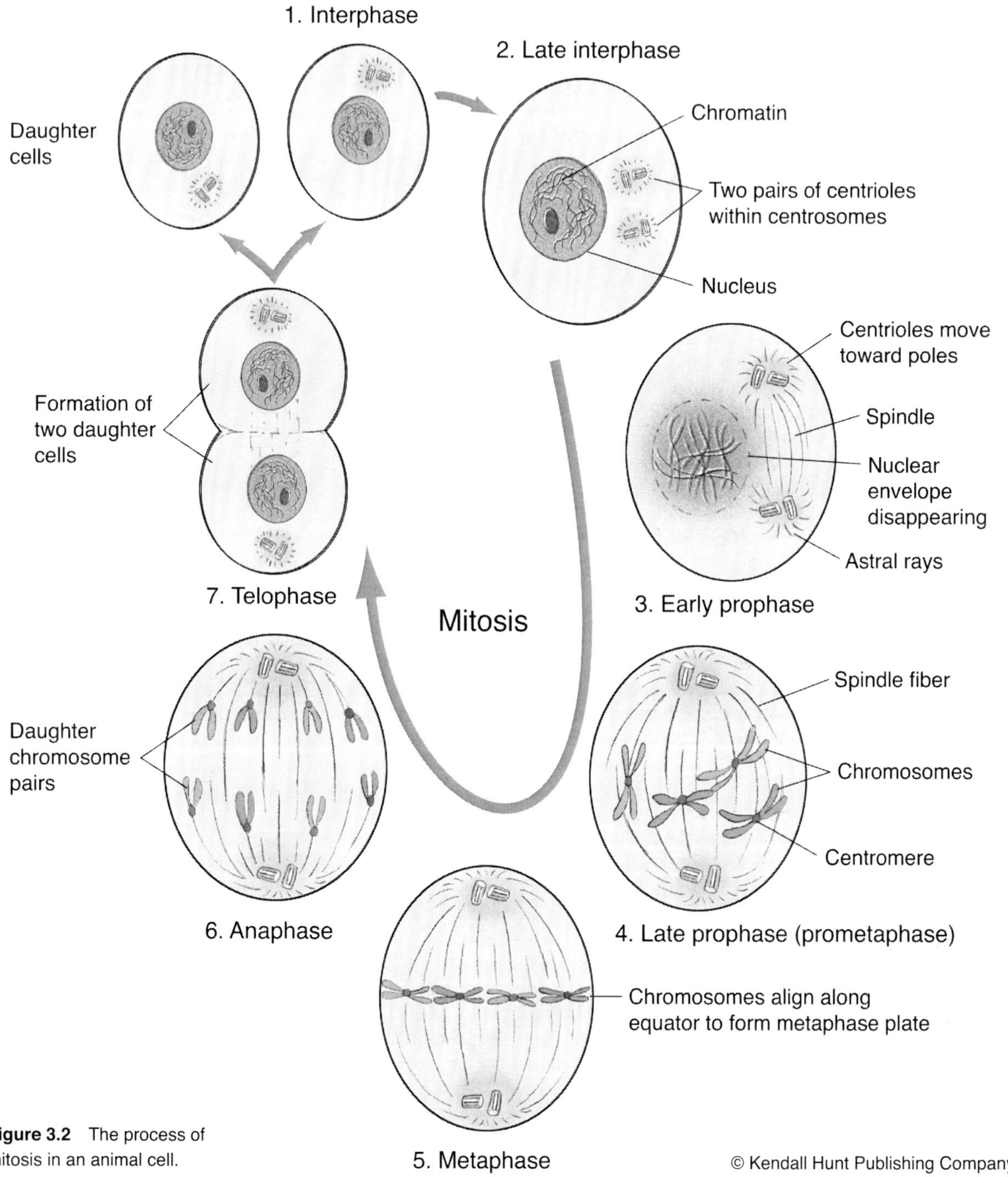

Figure 3.2 The process of mitosis in an animal cell.

Meiosis: Evolution's Revolution

It should be apparent from the preceding discussion on mitosis that asexual reproduction leads to genetic uniformity. In other words, at the conclusion of mitosis, two genetically identical daughter cells are produced. Thus, huge populations of identical cells, whether tissues or amoebas, can be issued in a relatively short period of time. But uniformity, as important as it can be, is not the result, nor is it the goal, of sexual reproduction. On the contrary, sexual reproduction leads, not to homogeneity, but to genetic variability. Diversity is its very essence. And keep in mind that it's sexual reproduction that's chosen by the vast majority of organisms.

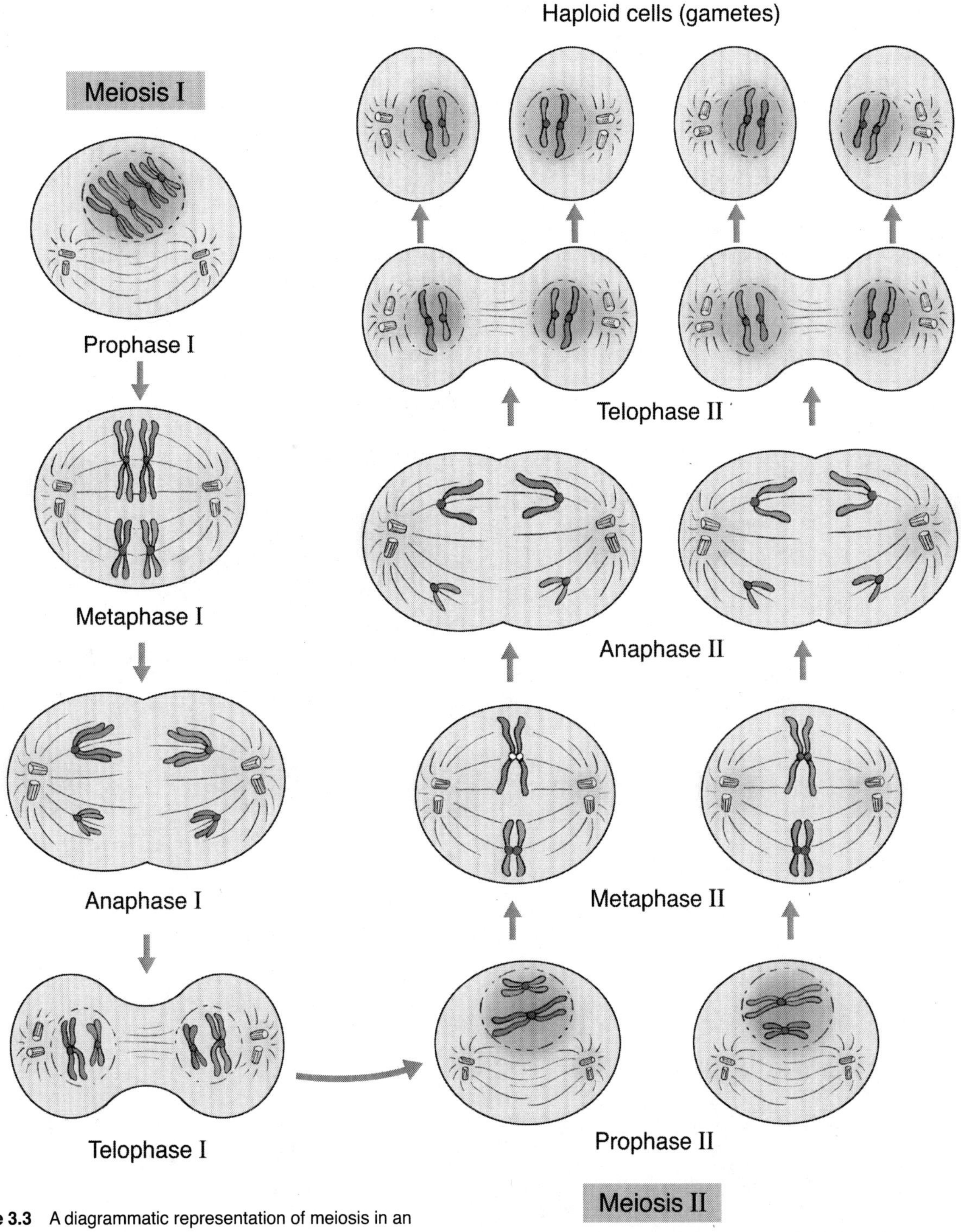

Figure 3.3 A diagrammatic representation of meiosis in an animal cell.

While we may think of sexual reproduction as a process requiring two separate parents, that's not always the case. Indeed, hermaphroditism is alive and well in the animal world (see Topic 4). The key to understanding the sexual pattern of reproduction is not in the action of separate parents, but in the production and action of two different sex cells called **gametes**. These two cells, each carrying different genetic information, unite during fertilization to form a new, genetically unique, cell, the **zygote**. *Sexual reproduction then is the union of different sex cells, usually termed sperm and egg, and usually, but not always, derived from a male and a female parent.*

Since the gametes are genetically distinct from each other, so then is the zygote a genetically distinct individual. That's the key to sexual reproduction. Soon after its appearance, the zygote begins to divide into an ever-expanding explosion of cells...it begins to grow and develop. To do so, it relies upon mitosis. But at some stage, let's call it adulthood, the new individual has the capacity to contribute its genes to the production of gametes, sexual reproduction follows, a new (genetically distinct) zygote is formed to eventually reproduce, and so on. You can now see that the secret to sexual reproduction lies squarely in the production of gametes. And how are gametes formed? The answer is not found in mitosis, but through a similar process called meiosis.

Meiosis starts out in the same manner as mitosis in that chromosomes are replicated. But instead of the sister chromatids separating from each other, they stay together and each links with a predetermined partner, its **homolog**. The diagram on the previous page shows this process.

Homologs are defined as a pair of chromosomes carrying genes for the same traits: eye color, skin color, fingernail shape, and so on. These homologous chromosomes come together early in meiosis in an event called synapsis. Following synapsis, the homologs (still consisting of two sister chromatids) are separated from each other and distributed to two individual cells. Each of these cells then undergoes a process reminiscent of mitosis in which the sister chromatids do finally separate into the formation of two new cells. Therefore, as a result of meiosis, four daughter cells (gametes), each containing one-half the number of chromosomes of the parent cell, are produced. In other words, during meiosis a **diploid cell** undergoes reduction division to produce **haploid gametes**. Two of these gametes (sperm and egg), each with a different genetic pattern, unite during fertilization to produce a new, genetically distinct individual... the zygote.

Of utmost importance in sexual reproduction is that each newly produced individual is genetically different from every other member of that species. The lone exception is when identical twins develop. Other than that, all sexually reproducing organisms are distinct.

The importance of this diversity is difficult to overstate since the entire evolutionary process in higher animals depends upon it. Because members of a population differ from each other, some will most likely have a competitive advantage over others of their race; they are better fit for their ecological niche. They will have a higher survival rate. Those that survive, or otherwise out-compete their cohorts, will, of course, be the ones to reproduce. The result is that the survival genes are passed along. (see Topic 16)

IT'S A FACT

WHY DO THE VAST MAJORITY OF ORGANISMS "CHOOSE" THE SEXUAL PATHWAY? THE ANSWER IS CLEAR. THEY DO SO BECAUSE BIODIVERSITY IS THE ROUTE TO EVOLUTIONARY SUCCESS. AND WHAT'S RESPONSIBLE FOR THIS GREAT DIVERSITY? WHY, NOTHING LESS THAN THE INTERMINGLING OF GENES RESULTING FROM MEIOSIS.

Topic 4

Animal Reproduction

Reproduction is one of the most important events, if not the most important event, in the life of an animal. In certain respects we can view an individual animal as just a storehouse for the genetic instructions wound within the DNA molecule. Each animal temporarily holds a small share of the total genes (i.e., gene pool) for its particular species to be passed on to the next generation. Members of that next generation then store the message for a short time only to convey it on to their heirs, and so on. Certainly, from the perspective of the population (i.e., the species), it is absolutely crucial that an animal passes along that genetic baton; chastity, for all of its supposed merits in the world of humans, goes unrewarded in the rest of the animal kingdom. Reproduction, then, can be thought of as the *measure of success* within a population. Among animals, there are but two methods available—asexual and sexual.

Asexual Reproduction

Asexual reproduction refers to multiplying through mitosis: one cell producing two identical genetic copies of itself. Many simple organisms reproduce primarily through mitotic division. In unicellular organisms (e.g., amoebas, flagellates, etc.), the one cell divides into two, a process termed binary fission. In another form of asexual reproduction, **multiple fission**, the nucleus of a single cell divides simultaneously into a number of individual nuclei, resulting in large numbers of new cells. In some cases of multiple fission, such as **sporogony** and **schizogony**, the number of new cells produced can be utterly astounding as witnessed in the malarial parasite, *Plasmodium.*

Other patterns of asexual reproduction include **budding**, **fragmentation**, and **gemmule** formation. Many multicellular, and a few unicellular, animals reproduce asexually through a small outgrowth from the body of the adult. In this type of reproduction, called budding, the new growth attains a critical size then drops off to assume an independent life of its own. Budding is widely used by ciliates, sponges, many cnidarians (e.g., *Hydra,* sea anemones, and *Obelia*), and even in some annelid worms. Another form of asexual reproduction is **fragmentation**. Here pieces of the adult break away to continue growth into a new individual. Sponges, sea anemones, and even starfish can reproduce in this manner. The final example of asexual reproduction is illustrated by sponges. Many freshwater sponges cannot withstand freezing or drying conditions and so have developed a strategy to bypass adverse circumstances. Special capsules called **gemmules**, containing amoeboid cells, are highly resistant to low temperature and desiccation (i.e., drying out) and are thus able to over-winter safely. Upon resurgence of favorable conditions, the amoeboid cells emerge to become the next generation of sponges.

Sexual Reproduction

Although many animals use asexual reproduction in one or more of its forms, the vast majority choose, instead, to multiply through sexual means. Here *meiosis,* not mitosis, is the mechanism. As a result, haploid sex cells called gametes (i.e., spermatozoa and ova [eggs]) are produced. Each gamete carries specific genetic information, usually (but not always) from a different diploid parent. Once the two gametes unite during fertilization, a new, genetically unique, individual is formed. In other words, unless the product of identical twinning, each offspring produced sexually possesses a combination of genes that makes it a unique being, different from all others of its race. Although the patterns and structures involved in sexual reproduction are many, there are only two basic pathways.

(1) **Hermaphroditism.** In many animals a single individual has the capacity to produce both sperm and eggs; it's a hermaphrodite. If an animal can form both egg *and* sperm, it only follows then that it might have the ability to fertilize

itself. In actuality, some animals do just that, especially when they live a solitary existence. For example, tapeworms often occur singly within an intestine. If they are to take advantage of sexual reproduction, they have little choice but to self-fertilize. Some sessile or attached animals, too, are self-fertilizers. They may rarely encounter another of their kind, so fertilizing one's own eggs is the only sexual outlet.

However, it seems that if given the opportunity tapeworms and other hermaphrodites always elect to use two separate individuals to reproduce sexually. They apparently prefer mating partners. In such cases, when an egg from one individual is fertilized by the sperm of another, **cross-fertilization** has been achieved. Keep in mind that when this occurs, genetic material (usually sperm) is exchanged between both hermaphroditic individuals, and so each animal is simultaneously acting as both male *and* female. We call this type of synchronized pairing **mutual copulation** or, as it is sometimes referred, **simultaneous hermaphroditism**.

(2) **Separate sexes.** As we move along the parade of animals, we speak more and more about males and females. The reason for this is that hermaphroditism gradually gives way to separate sexes. Here, an animal can make either eggs or sperm, but not both. In general, gametes are produced by specialized organs called **gonads**; males produce sperm in testes, while eggs are produced by females in ovaries. *This strategy is used by the great majority of animals; most invertebrates and all vertebrates.*

For sexual reproduction to be successful, sperm and eggs must come together in the act of fertilization. There are only two ways this can occur: **external fertilization** and **internal fertilization**. In many groups (sponges, corals, annelids, various fishes), gametes are merely released into the surrounding water where external fertilization occurs. In **broadcast spawning** of this nature, huge numbers of eggs and sperm are discharged with the prospects that some will unite and survive to adulthood. Parents provide little if any care of the young in this reproductive strategy.

As you can imagine, broadcast spawning has some serious drawbacks. Too much is left to chance; egg and sperm may never collide, the aquatic environment may be hostile, and predators may eliminate most, if not all, of the offspring. The process is just too hap-hazard. So, to counter the gamble inherent in external fertilization, many aquatic and all terrestrial animals have opted for *internal* fertilization. Here, sperm is actually introduced into the female where fertilization is all but assured. To be effective, however, two things are required. First, the two animals must come together. This, too, has some risk. Qualified, mature mates may be hard to find, females may already be impregnated, or they may just be unreceptive. Thus, we see arise a whole array of behaviors or **courtship rituals** designed to bring the two sexes together. In biology, there is no amphitheater more bizarre or fascinating than that of the courtship arena. The lengths to which some animals will go to establish mating bonds is truly amazing. From insects to birds and mammals, the time, energy, effort, and evolutionary creativity of courtship can be utterly astounding.

The second problem encountered with internal fertilization is one of **copulation**. Once the two sexes have gotten together, sperm must be introduced into the female. To ensure that this happens in the most effective manner, a number of male structures or devices have evolved. These intromittent organs, all designed for the delivery of sperm, include stylets, modified appendages, palps, tentacles, cirri, gonopods, and penises.

Special Situations

There are two reproductive schemes in animals, conjugation and parthenogenesis, that defy easy classification. Each seems to lie somewhere between the asexual and sexual strategies.

Conjugation occurs in some ciliates (e.g., *Paramecium*), and although it seems to be a sexual process, it isn't. Separate sexes do not occur, and, in fact, sperm and eggs are not even formed. Instead, two mating types line up side-by-side and remain attached for several hours. During the pairing process each individual undergoes meiosis to produce haploid pro-nuclei, which are exchanged with its partner. Following this exchange of genetic material, the two organisms separate and each undergoes two sessions of mitosis to produce four individuals. Note that in conjugation there are no sperm or eggs per se, yet new genetic combinations do appear because of meiosis.

It is relatively common for a number of otherwise sexual animals (flatworms, roundworms, crustacea, and insects) to reproduce without normal fertilization. **Parthenogenesis** (Greek partheno, "virgin"; genesis,

Figure 4.1 The dioecious condition in animals, in which separate sexes emerged, gradually gave rise to internal fertilization. This, in turn, gave rise to courtship rituals and eventually to copulation between the sexes as shown by the mating beetles and leopards.

"birth") is the term given to a process whereby an embryo develops *without* input from both sperm and egg. There are several species of fish, for example, that reproduce parthenogenetically as well as a number of amphibians, lizards, and snakes. It has even been recorded within domestic turkeys.

Sometimes sperm are marginally involved and serve as a mere stimulus toward egg development but do not contribute their chromosomes to the process. In most instances of parthenogenesis, however, the male is even less involved; he's superfluous and doesn't participate in any way. In fact, in some populations there are no males at all! How is that possible? Well, most female animals have identical sex chromosomes (e.g., the familiar **xx** pattern of the human female). If an **xx** egg develops parthenogenetically (without spermatic input), then the only possible outcome is **xx**, or females. These species, in fact, consist of nothing but females.

As a reproductive pathway, parthenogenesis seems to have elements of both asexual and sexual reproduction. It appears to be asexual in that only one "parent" is involved. Yet, at the same time, it is somewhat sexual in that meiosis is used to form the eggs and so the offspring may be slightly different, genetically, from their mother. Entire generations of certain animals, such as aphids, ants, bees, fish, amphibians, and reptiles, may reproduce exclusively in this rather odd fashion.

However, when viewed from a broad perspective, a definite trend emerges—reproduction evolves from asexual to sexual. The fission of single-celled organisms yields to budding and/or fragmentation. Next emerges those species that use a combination of budding/fragmentation and hermaphroditism. Finally, separate sexes are attained. And once that happens, there's no turning back. In the vast majority of animals, starting with the nematodes (i.e., roundworms), sperm are formed only by males, and eggs, in turn, by females. This is evolution's choice for virtually all species of mollusks, arthropods, and chordates (including fishes, amphibians, reptiles, birds, and mammals).

Asexual or Sexual... Which Is Better?

At this point you may very well be wondering why all the different reproductive patterns? Or you may be asking whether there's any reason for an animal to select one pattern over another? Is it better to reproduce asexually or sexually? Are there advantages that one strategy offers in comparison to another? The answer to these questions is that there are advantages to both, depending upon circumstances.

The Asexual Arsenal

When we examine the asexual strategy, we uncover two important factors: (1) asexual reproduction is practiced by the more simple animals, primarily protists and certain "lower" invertebrates, and (2) the vast majority of asexual animals are aquatic, especially marine. Putting these two factors together gives us real insight into the appeal of asexual reproduction. Many habitats within aquatic environments are relatively stable; they change little. If a species is already well tuned (adapted) to its aquatic habitat, it's to that population's advantage to pass on those very fitness genes to the next generation. In other words, *genetic uniformity is an advantage in a stable environment.* Since asexual reproduction is based upon the process of mitosis, all offspring are genetically identical to each other. Thus, generation after generation the population remains suited to its aquatic home. In addition, the process of asexual reproduction, since it doesn't require a partner, is easy and quick. This is an important consideration for the more simple animals that may lack the sophistication to form gametes or to locate mates. Thus, energy and time is not wasted seeking a partner, convincing that partner of your desirability, and engaging in copulation.

A final, but very important, advantage of asexual reproduction is that it can produce huge numbers in a relatively short period of time. The numbers can be simply overwhelming.

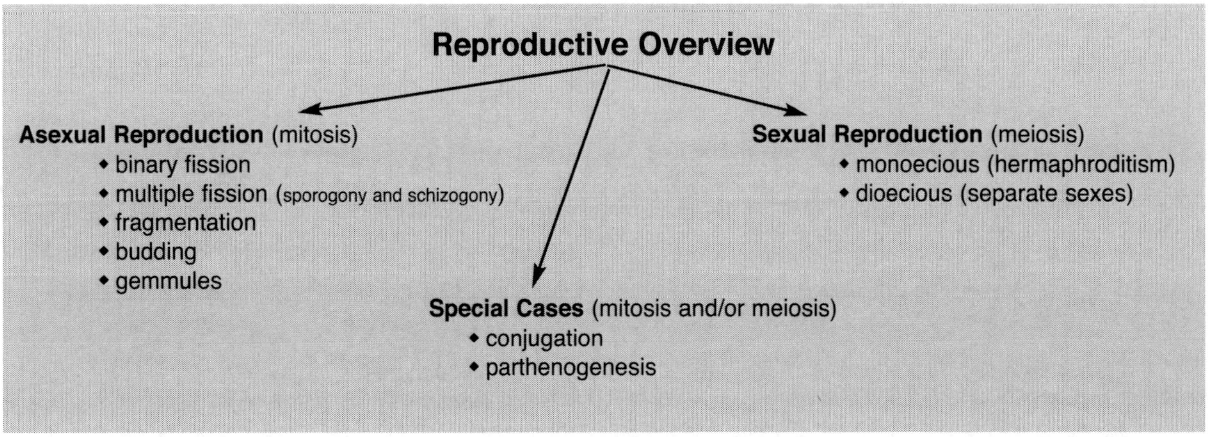

THE ALLURE THEN OF ASEXUAL REPRODUCTION IS THAT IT'S GENETICALLY UNIFORM, EASY, FAST, AND PROLIFIC.

On the other hand, environments are not always static; they are subject to change, say due to drought, a severe fluctuation in temperature, or even change brought on by the invasion of a virulent disease. If that happens, then the very advantage of asexual reproduction may become a population's downfall. If a well-adapted, genetically uniform asexual species suffers such a catastrophe, it most likely lacks the genetic resources to survive the onslaught. Why is that? To understand the problem here, we must realize that natural selection operates upon genetic variety in a population. When variety is lacking, as it is in asexual populations, the entire population may fall prey to a catastrophic change. An asexual population's only source of genetic variation is through mutation. However, mutations are unpredictable and are, in any case, primarily detrimental or lethal. So mutations in asexual populations are just too unreliable to offset a drastically changing circumstance.

ASEXUAL REPRODUCTION'S MAJOR DISADVANTAGE LIES IN ITS VERY GENETIC UNIFORMITY. IT JUST DOESN'T WORK WELL IN A CHANGING ENVIRONMENT.

The Sexual Invasion

The vast majority of plants and animals that inhabit the earth are sexual. This tells us that sexual reproduction must hold a powerful attraction. And, as with asexual reproduction, the reason is linked to the environment. Most animals live in habitats that undergo periodic, and at times, regular change. As indicated in the discussion above, a population's survivability during environmental insults is related to genetic variation. The more diverse the population, the more likely some individuals will be preadapted in favor of a change. As a result, they are the ones who will survive and reproduce to pass along their survival genes. But sexual reproduction has several drawbacks: (1) in many cases, it's an energy-demanding process; (2) it's often metabolically wasteful in that many eggs and sperm go unused; and (3) it may be time-consuming (locating a proper partner, persuading that partner to accept you, and so on). Nonetheless, sexual reproduction has one mighty advantage...**genetic variation**.

Sexual reproduction relies on the formation of gametes produced through meiosis. When fertilization results, a new, genetically unique individual is formed. It is upon this variation that natural selection can exercise her magic wand. Evolution in a changing environment depends mightily upon this genetic variety. Whereas asexual reproduction preserves the genetic *status quo* through mitosis, sexual reproduction, by its very nature of mixing genes from both gametes, dismantles it. But by dismantling the parental genotype, new and unique genotype are created. The effect of this inherited variability is so potent that most animals have opted for male-female reproduction. The clear advantage of separate sexes is the genetic versatility bestowed by the combination of genes from two distinct parents.

IT'S A FACT

THE OVERWHELMING POWER OF SEXUAL REPRODUCTION IS ITS GENETIC VARIETY WHICH PROPELS EVOLUTION WITHIN CHANGING ENVIRONMENTS. THIS MORE THAN COMPENSATES FOR THE ENERGY EXPENDED, THE WASTEFULNESS, AND THE DEMANDS INHERENT IN THE SEXUAL PROCESS.

The Pros and Cons of Reproductive Strategies

Asexual Reproduction

Disadvantages = nonadaptive; cannot readily adapt to changing environments

Advantages = genetically uniform, quick, easy, and prolific; works well in a stable environment

Sexual Reproduction

Disadvantages = time-consuming, energy demanding, may waste gametes

Advantage = genetic variation permits adaptation to changing conditions.

Topic 5

What's in a Name?

A Question of Relationships

There are millions of organisms occupying planet earth. And while each is unique, different from all others, there are, nonetheless, definite patterns of similarity. Evolution has produced some animals that are very much alike and yet are quite different from others. It is obvious that animals with feathers are more closely related to each other than they are to animals adorned with fur, or with a scaly covering, or to those with slime-covered bodies. One-celled organisms are more closely aligned to each other than to multi-cellular forms. Roundworms are quite different from jellyfish. It is upon these shared relationships that scientists have developed an organizational scheme…a classification system to help us grasp the order in life's vast diversity. What may seem like a chaotic, disorderly sea of organisms becomes sensible and manageable when we apply a systematic approach.

There are two disciplines that deal with life's vast diversity. **Systematics** is the study of the relationships that occur among organisms and their evolutionary history (phylogeny) while **taxonomy** is the actual naming and classifying of organisms. As a result of these two activities, each known organism is placed into a taxonomic hierarchy that reflects its relationship to all other organisms. The hierarchy is based upon the early work of Karl von Linnaeus (1707–1778) and is known as the Linnaean classification system. He realized that organisms could be placed into groups based upon shared relationships. As a result he devised a ranking system consisting of the seven distinct levels listed below. Note that each level (called a taxon) is more inclusive than the one below it.

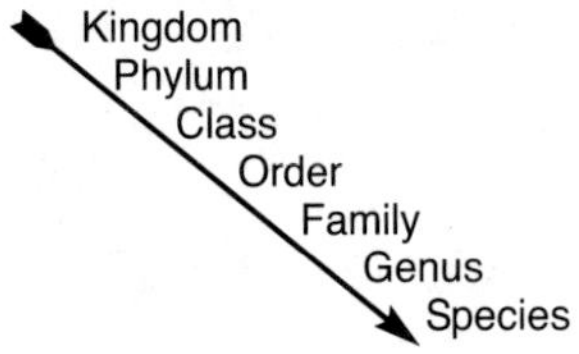

Although early scientists recognized only two kingdoms (plants and animals), we now distinguish five:

Monera* prokaryotic cells (bacteria and cyanobacteria)

Protista eukaryotic, single-celled organisms

Fungi multicellular, saprophytic decomposers (slime molds, yeast, mushrooms)

Plantae multicellular, photosynthetic organisms (plants)

Animalia multicellular, heterotrophic organisms (animals)

*Some biologists have divided the monerans into two separate groups, the Archaebacteria and Eubacteria, giving us six kingdoms. We, however, will stay with the more traditional system of five.

The Genius of Linnaeus

In addition to giving us a logical system to categorize all organisms, Linnaeus also devised a way of naming each and every one. It's a brilliant, ingenious approach that gives each animal a name consisting of two parts. The first part is the **genus** name and the second is the **specific** (species) epithet. Together they make up the scientific name of an organism that's unique to that one species. In this way much of the confusion inherent in common names is eliminated. Take, for instance, the largest carnivore in the United States, the mountain lion. In various parts of the country it is known as the cougar, panther, painter, and puma…all the same animal. Also, it may be confused in the southwestern part of its range with the jaguar, or in the West with the bobcat. However, by using its scientific name, *Felis concolor,* biologists from across the country can be sure they are referring to the same animal. That's the beauty of the **binomial system**. It's unique for every species; it creates order out of chaos.

Rules of Binomial Nomenclature

When a new plant or animal is discovered it's eventually given a name. To do this, a number of steps, or guidelines, must be followed. The more common ones are listed below:

(1) Each scientific name has two parts: the genus and the species epithet. For example, *Amphispiza belli* is the sage sparrow. Sometimes a third name, the subspecies, is used for distinct subpopulations. For example, *Amphispiza belli belli,* is the San Clemente Island sage sparrow that occurs only on that island.

(2) The name is to be in Latin or otherwise "latinized," such as *Crotalus horridus,* the timber rattlesnake. Note that the specific name is an adjective that often describes a characteristic of the animal. What feature do you think the spotted skunk, *Spilogale putorius* has, or how about the rabbit *Sylvilagus aquaticus*?

(3) The name is to be set off in different typeface, usually in italics, or underlined.

(4) The first letter of the genus name is to be in uppercase; all others, including the first letter of the species name is written in lowercase.

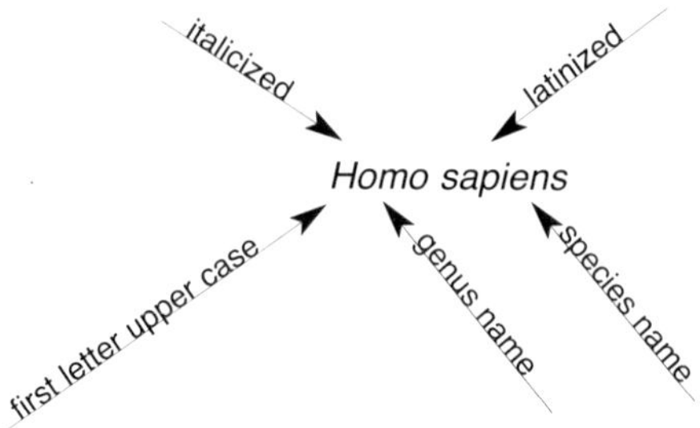

Biodiversity

Introduction

That animals are spectacular is a conclusion often reached by the most casual observer. Indeed, even experienced biologists find their study inspiring. And why not? When you consider the many ways animals differ from each other—their enormous size variations, the contrasting palette of colors, their bizarre structures, the changing complexion of behaviors, the cacophony of sounds, and their mosaic habitats—you begin to understand the awe-inspiring nature of animals.

At first glance this world may seem hopelessly confusing: just a hectic collection of mysterious and bewildering creatures running, swimming, crawling, and flying about aimlessly. There seems to be little design…just randomness. However, a closer look reveals not a world of chaos, but one instead interlaced with structure, organization, intricate patterns, and complex relationships. An underlying order exists.

It is this order, this hidden inner sanctum, that biologists struggle to uncover. Through observation and penetrating investigations, we are slowly unmasking the secret universe of animals. In fact, for some of the more common species we have already amassed a fairly large reservoir of knowledge concerning their structures, needs, habitats, and behavior. For instance, there has been more than one book written about the common housefly or the cockroach. Certain mosquitoes are well known. The common earthworm, the bald eagle, and the white-tailed deer are also some of the better-known species. But for the vast majority of animals we have barely scratched the surface. We know virtually nothing of their anatomy, of how or where they live, of how they interact with each other. For them our encyclopedia of understanding is appallingly empty.

How Many Kinds?

One of the very first questions people ask about animals has to do with their numbers. Animals seem to be all around us, so exactly how many kinds are there? To find that answer we look to **taxonomy**, the science of naming and classifying organisms. When we do, we get a mixed picture. First of all, we learn that about 1.2 million different animal species have been discovered and named so far. However, the real question is not how many animals we have met face-to-face, but instead, how many kinds actually exist? And here the picture gets somewhat muddled. That's because our best estimates tell us that the one million or so thus far identified is just a small fraction of the total. But what that total really is we're just not sure. Oh, we can offer educated guesses, but they are just that…guesses. If you were to ask a flock of zoologists how many species of animals exist, you would surely receive a flutter of answers. On the more conservative side the number may be somewhere in the neighborhood of 10 million species. However, most experts would place that number much higher, somewhere in the 40 to 50 million species range…maybe even twice that number!

Whatever the exact number might be, and we will never know for sure, it at least means that there are millions and millions of animals scrambling about, most unknown to us. Each is surviving in its own little corner, differing from all others in its basic needs, anatomy, physiology, specialization, size, behavior, and so on. Each has found a special place, a special role, a special way in which to thrive. Think about it... animals are everywhere: under rocks, on tree trunks, hidden in decomposing carcasses, clustered amid vegetation, teeming in rivers, lakes, and oceans, swarming in the air, abundant in their secret underground labyrinths. They even persist in deserts and polar barrens. There are thousands in each handful of humus, in each cup of pond water. Swamp muck is rich with them; estuaries are flooded with their heirs; the rain forests are showered with their uniqueness. The earth literally pulsates with their activity. Whether you prefer to use a microscope or a telescope, you will find them everywhere.

How Many Individuals?

Perhaps as important as the number of different species is the number of individual organisms lurking about. Just how many robins are there? How many houseflies, wildebeest, cockroaches, bees, or sidewinder rattlesnakes. How many fleas, lice, or mites? How many soldiers march within the vast legions of army ants? How many sharks police the oceans? How many termite workers toil in their subterranean bivouacs? What is the number of earthworms plowing the soil or seahorses riding the waves? The number of roundworms alone is said to be in the millions in each square meter of soil. A single gulp of planktonic brew by a megamouth shark may engulf billions upon billions of individual, microscopic animals.

So what is the total number? Just how many individuals are alive at any one time? Plainly, it's not possible to know. There are too many to count. They move too fast, are too secretive, hatch too quickly, and die too rapidly: a census is clearly out of the question. Yet we can imagine in our mind's eye what that massive pyramid of animals is like. And from that mental image, we realize that their number must be staggering, their biomass enormous, and their impact immeasurable!

Understanding Diversity

While we cannot be sure of how many different animals exist nor of the total number of animals on earth, we can agree that the sheer volume is mind-numbing. But what do the numbers have to do with biodiversity? To answer this question we must first get a handle on the meaning of the term diversity. In its simplest context, biodiversity refers to the kinds of organisms living in a given area. That area may be as small as a rotting log, a pond, or the canopy of a single tree. On the other hand, the area may encompass ecosystems as vast as deserts, forests, prairies, oceans, or the entire biosphere. Regardless of the area, *diversity is the kinds of different organisms living there.*

In general, we think of diversity as the way in which species vary from each other: an alligator differs from a woodpecker, which differs from a butterfly, which, in turn, differs from a tapeworm. This is **interspecific diversity**...each species is different from every other kind. There are even different types of woodpeckers and different types of butterflies. But this, too, is interspecific diversity because each kind of woodpecker still belongs to a separate species: downy, red-headed, red-bellied, three-toed, and yellow-bellied sapsuckers...all woodpeckers, but each belonging to a different species. And there are literally thousands of different species of butterflies.

Yet diversity can be expressed in another way. Each member of a given species, especially in higher organisms, also differs from all other members of the same species. Each red-headed woodpecker is different from all other red-headed woodpeckers. Every monarch butterfly varies from all other monarchs. Just think of how we humans are each unique. Unless you are an identical twin, you are different from every other person alive. This is the second level of diversity, **intraspecific diversity**; variety within a given species. Each member of a sexually reproducing species differs genetically from all other members of that same species.

When biologists use the term diversity they are usually referring to the different species that may occur in an area, not to individual raccoons or paper wasps.

Figure 6.1. Each species is adapted to a specific niche where it secures the needed resources for life. For example, a moray eel (above) lies in wait of prey along the ocean bottom while a sea otter (bottom) prefers to bring its food to the surface.

Figure 6.2. All members of a sexually reproducing species differ genetically from each other as seen here in an assemblage of monarch butterflies.

In other words, they mean *interspecific* rather than intraspecific diversity. We are, in most cases, exploring how animal species differ from each other in their needs, structure, function, and behavior.

So, What Is a Species?

Both interspecific and intraspecific dynamics are interwoven to give us the tremendous variety of animals on planet earth. One depends upon the other. But before we explore just how this occurs, let's first see what is meant by the term **species**. What criteria are used to differentiate one kind of animal from another? How can we tell them apart? We know from personal experience that sometimes it's quite easy. For example, discriminating between an elephant and a fish is simple, and distinguishing a robin from a hawk is not so difficult either. On the other hand, separating some species from others can be nearly impossible. Beetles, belonging to the insect taxon Coleoptera, are perhaps the most numerous kind of animal on earth. There are literally hundreds of thousands (maybe millions) of different kinds of beetles. How can anyone sort out one beetle type from another? Can anyone tell slugs apart? Can you distinguish between the several kinds of "sea gulls" that inhabit our coasts or lake fronts?

The task can be daunting. Yet it's highly meaningful to know what a species is and how to tell one from another. Otherwise, we cannot logically discuss animal behavior, habitats, breeding biology, and so on. So we must search for a definition, but finding a good one is difficult. However, one definition that seems to satisfy many biologists is that a species is *a group of organisms sharing a common gene pool.* In other words, if two animals mate and yield fertile offspring, who, in turn, can also successfully reproduce, then both are, according to this **biological-species concept**, members of the same species. If they can't, or don't, they belong to separate ones. So if *animals are reproductively incompatible* (i.e., cannot produce fertile offspring), *they exist in nature as entirely different species.*

Stated in another way, this definition declares a species to be a population of organisms reproductively isolated from all other populations. They are separated in such a way that they cannot or do not reproduce; they are reproductively isolated. And the actual isolating mechanisms can be wide ranging. They may consist of **geographical separation** in which two populations never come into contact with each other. Or isolation may be **behavioral** (i.e., courtship rituals are incompatible), **physiological** (i.e., the egg and sperm cannot mix), or even **anatomical** (i.e., reproductive structures are mismatched). In any event, the result is always the same: mating cannot or does not yield viable offspring. Thus, their genes do not intermingle. The available genes within each population (i.e., the **gene pool**) remain intact.

As reliable as this definition might be, it's not without one limitation. It works quite well for sexually reproducing organisms, those producing eggs and sperm and, thus, usually requiring a male and a female organism. But the definition does not apply to asexually reproducing animals. An asexual animal doesn't require a mate, it doesn't exchange genes with another animal, and therefore the question of reproductive compatibility is pointless. For them...and there are large numbers of asexual animals...a definition of species relies more upon their phenotypes. In other words a definition of species for asexual animals is *a group of similar organisms that are recognizably different from all other groups.*

> FOR THE MAJORITY OF ANIMALS, A SPECIES IS DEFINED AS "A GROUP OF SIMILAR ORGANISMS REPRODUCTIVELY ISOLATED FROM ALL OTHER GROUPS".

If there are indeed 10 or so million different species of animals on earth...why so many? And how did they get to be so numerous? To address these questions, we only have to look to the daily life of individual animals. When we examine them carefully, we discover that each one shares two things with all others. Regardless of the animal in question, it is preoccupied with two simple strategies: *staying alive* and *reproducing.*

Self-Preservation

Animals do whatever they can to stay alive. They struggle daily to do so. In fact, a large portion of their time and energy is spent doing just that. They must eat to survive and at the same time be sure that they do not become someone else's dinner. They need to hunt without becoming the hunted. They use what sensory structures they possess to become aware of their surroundings so that food is found and enemies avoided. They scurry about finding food and shelter and all the while staying out of harms way. This is all part of staying alive.

However, self-preservation goes beyond just food and safety. Staying alive also involves maintaining inner stability. **Homeostasis** is the balancing of internal processes so that an animal functions within a narrow range of efficiency. Such things as proper temperature, ion and fluid balance, pH level, and scores of other metabolic processes must be continually monitored and adjusted. These homeostatic mechanisms, too, demand their share of an animal's energy budget.

There's no question that an animal spends much time, effort, and talent on just staying alive, on surviving. But, it is also true that, at times, just as much of its energy, perhaps even more, is expended on reproduction—on **self-perpetuation**. In point of fact, we can argue that animals make such a dedicated effort to stay alive just so they can reproduce. Reproduction is that strong of a drive. Does that mean that the very "goal" of life then is procreation? That question is one of the most important in biology. Let's see if we can find an answer.

Figure 6.3. Animals are consumed by two preoccupations: self-preservation (staying alive) and species perpetuation (reproducing).

Multiplying

Biologists realize that animals have a compelling urge to reproduce and can do so with alarming efficiency. In fact, animals, at times, can increase far beyond our imagination. Unchecked, their numbers can skyrocket to unbelievable proportions. Examples can be drawn from a variety of sources: insects, such as grasshoppers, locusts, ants, and flies can produce huge, swarming populations in short order; squid do likewise; mice and rats, too. Flamingos take flight in immense pink clouds numbering well over a million individuals. Passenger pigeons traveled in such huge flocks that they eclipsed the sun. Even penguins occur in enormous breeding colonies. A single female human roundworm (*Ascaris*) located in your small intestine will carry in excess of 27 million eggs in her uterus at any given time. She releases, on average, about 200,000 of these each day. That's something like 73 million eggs in one year. However, a single oyster may hold the record; it can produce over 500 million eggs in a single year!

However, one doesn't need to be a biologist to realize that animals have an unquenchable desire to reproduce. We have all heard of the suffocating plagues of locusts or the swarming, marching columns of army ants. We have all suffered the mass emergence of gnats, mosquitoes, bees, or wasps. We have, indeed, all witnessed the result of this ravenous need of animals to propagate.

Truly, animals are great reproducers and they spend an inordinate amount of energy and time doing just that. In fact, occasionally the whole procreative process appears chaotic, confusing, and even counterproductive. For instance, some animals regularly reproduce at great personal risk. Consider the many males who stake their very lives on the chance to court and mate: scorpions, tarantulas, black widows, and praying mantids, to name a few. Females of these species are cannibalistic and, following the mating act, will devour the male if given the chance. He may, if not careful, end up as his partner's post-nuptial snack!

Even the robin or cardinal in your backyard makes extreme sacrifices to produce next year's yield. They must establish and defend a territory, attract a mate, create pair bonds, build nests, lay and incubate eggs, and then are obliged to feed the relentlessly hungry nestlings. All of this is conducted at a substantial energy cost to the adults. And this scenario is repeated by countless animals during each breeding cycle.

When we think of what it takes for animals to reproduce successfully, it becomes obvious that reproduction is one of the most important undertakings in an animal's life. The process is, at the very least, an energy-demanding one, and at most, a survival risk. If that is so, then we must ask the simple question...why? Why does an individual animal spend so much time, effort, and sacrifice on reproduction? Why would a male face possible death just to multiply? What forces are there that compel them to do so?

The Urge to Reproduce

We know that the urge to reproduce is powerful and universal among animals. If that is so, then we can also assume that there must be some reason, some reward, some payback. But what kind of payback? Perhaps it just feels good; maybe animals reproduce in order to reach a state of procreative bliss. Do they march to the seductive drum of reproduction because of tender feelings? Hardly. It is rather difficult to imagine that organisms such

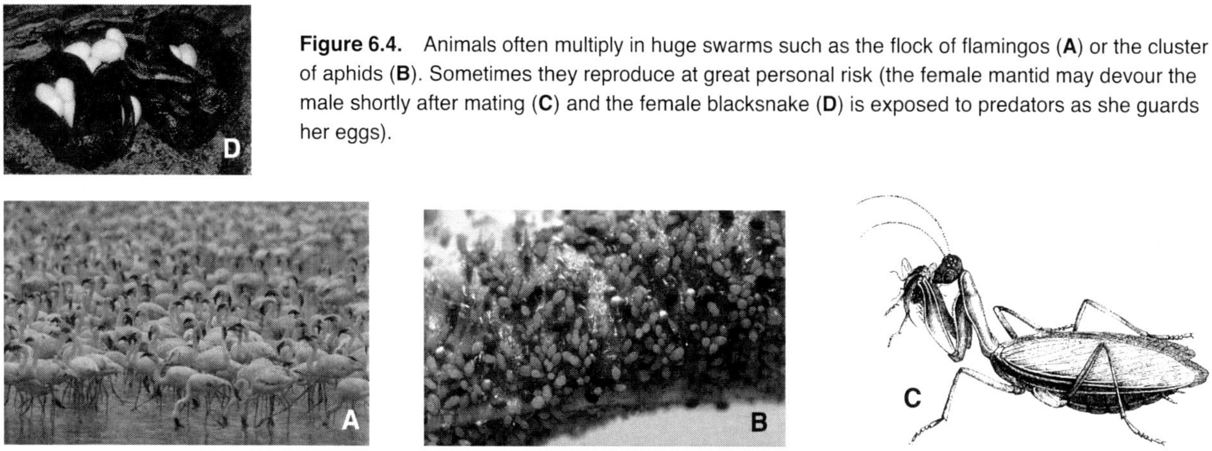

Figure 6.4. Animals often multiply in huge swarms such as the flock of flamingos (**A**) or the cluster of aphids (**B**). Sometimes they reproduce at great personal risk (the female mantid may devour the male shortly after mating (**C**) and the female blacksnake (**D**) is exposed to predators as she guards her eggs).

as the blob-like amoebas, the glassy sponges, the gelatinous jellyfish, hungry tapeworms, or blood-thirsty mosquitoes and millions of others reproduce just to reach a state of euphoria. The limited sensory structures of these "lower" forms scarcely support such an idea. And even with more complex species (except, perhaps, humans), it is hard to conceive that "sexual bliss" is the dominating drive. No, we must consider other forces at work here. So let's look at possible reasons a little more closely.

Just why do animals pursue mating with such persistence? If it's not for sexual bliss, then perhaps it's done to satisfy some paternal instinct...the conscious desire to care for offspring. Many animals develop a strong and enduring bond between one or both parents and the offspring. But is that the reason they mate?

Indeed, close observations of many "higher" animals, such as birds and especially mammals, reveal many patterns of such bonding. However, just because bonds are formed doesn't mean animals do so with any insight or foreknowledge. It doesn't mean they know what they are doing. The ability to form such reasoned, emotional, and reflective decisions would require a highly complex and sophisticated nervous system; one that probably only humans, and maybe a few other mammals, possess. Our current knowledge of the brain tells us that, among animals, only humans can figure out just why they perform certain acts, and only humans know what the probable outcome of these acts will be. Thus, it seems likely that other animals are destined to reproduce without ever knowing why!

Figure 6.5. All animals are intimately concerned with staying alive and reproducing. Here several different organisms feed and reproduce on a single plant.

Figure 6.6. Many animals develop strong bonds between parent and offspring.

Genes...Our Key to Immortality

If that is true, then they must reproduce instinctively, innately. They must do so because of a ready-made, built-in program. And, of course, we all know exactly what that program is. It's our genetic code—our chromosomes, our DNA, our genes, our very inheritance.

So, if animals multiply because of this code, then the most plausible explanation of animal reproduction is that they do so, not out of any desire or understanding, not because of any insight or passion, but rather in response to genetic commands, a set of instructions given to them by their predecessors. And it's only through the unraveling of the DNA code that the universal instructions are rewoven into the fabric of each new generation. It is the legacy of our ancestors, one we most eagerly pass along each time we reproduce. That thread holds the vital information for survival. It is a coded manual for every aspect of an animal's life: its development, daily behavior, body design, defensive activities, food preferences, and, yes, even its reproductive choices. The information interwoven within the DNA molecule is the very blueprint for life itself.

So, again, why do animals reproduce? The answer is clear: to pass along that valuable, timeless, coded DNA message...and they do so without even realizing it. It's the echo of the ancient genetic call. They propagate because the code compels them to.

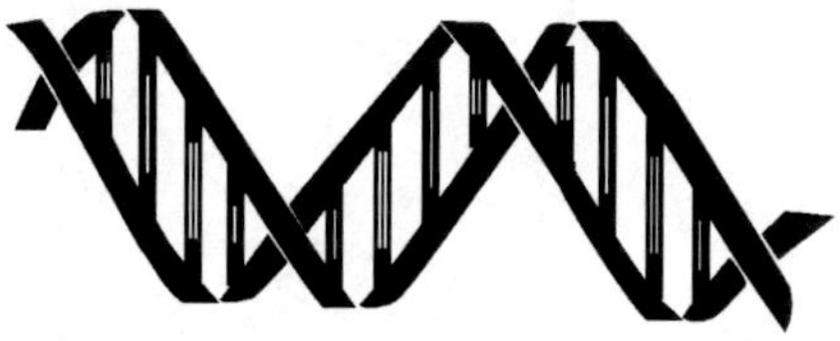

THE DNA MOLECULE IS A CODED, GENETIC MANUAL FOR DAILY BEHAVIOR, FOR REPRODUCTIVE CHOICES, FOR SURVIVAL ITSELF.

Topic 7

Just Who Reproduces?

This brings us to yet another intriguing question. Who reproduces? Exactly which members of a population get to hand down their genetic message? Is the whole process by chance alone, or is there some order, some kind of design involved?

Who reproduces? Why, survivors do, of course. Those very individuals who possess the capacity to flourish within their own habitats are most likely to pass on survival traits to their offspring, who will, in turn, survive to pass those same traits, and so on. The key point here is not that an animal survives, but that it survives long enough to reproduce, to pass on those survival skills to the next generation.

But is it through luck or by design? Well, there's no question that blind luck can play a part in who survives life's hazards and eventually reproduces. For instance, a nestling bird may fall into a river and drown or a gust of wind may sweep a spider far from its normal habitat before it has had time to meet a proper mate. But, on average, luck has little to do with who reproduces. Instead, propagation is primarily the outcome of an orderly, patterned process: an orchestrated genetic symphony played out on ancient, programmed instruments. An animal reproduces, *not because it's lucky, but because it has the devices and skills to survive long enough to do so.* This process is termed **differential reproduction**.

Figure 7.1 Reproduction is no haphazard process but is designed in a way to propel the best genes possible to the next generation. Here a single bull fur seal who has successfully defended his breeding rights guards his large harem.

Even If It Hurts?

We can now turn to the question of why some animals reproduce at great sacrifice. As part of the reproductive scheme, some males may be killed and eaten by the female immediately after they mate. This is true for many spiders, praying mantids, and a variety of other animals. We can envision how powerful the reproductive drive must be if an animal will face near-certain death just for the privilege. And we can further conclude that an animal must, therefore, have no idea of what he is doing. If he did, reproduction in that species would surely come to a screeching halt!

The strange thing is, this is not a rarity. In fact, throughout nature there are many instances of animals rushing headlong toward death in order to reproduce. Numerous species make the most extreme efforts to breed, often traveling hundreds of miles or waiting years to do so, only to die shortly thereafter. From squid to cicadas, from mayflies to salmon, entire populations of animals perish immediately after mating.

A well-known case is the pacific salmon. After spending most of its life at sea, it responds to the reproductive summons by swimming exhaustively upstream until it locates the very spawning grounds of its parents. There, where it hatched years before, a breeding orgy takes place. Following the release of sperm and eggs, all participants, both male and female, die by the thousands. Streams literally reek with the foul stench of their rotting remains.

A less celebrated, yet no less fascinating, example is the 17-year cicada. After almost two decades living underground, the immature stages, called nymphs, emerge in massive numbers and quickly molt into adults. Males then "sing" vigorously and incessantly to attract females. Mating takes place, eggs are laid, and then they all, too, soon die.

What does this tell us? Do you think these animals enter into this contest fully aware of what is about to happen? I hardly think so. So we return to our original question: "Why do animals clamber so to reproduce?" Clearly, the answer is that they simply don't know any better. They do so because of a genetic drive, an instinctive pulse to procreate at all cost. They reproduce because the process is etched within their genetic code. They are just following orders. And once again, who reproduces? It is those who are best suited for their particular nook. Why are they best suited? Because they received the success formula from the genes of their parents who were themselves well suited. The cycle continues through the generations by way of coded, genetic instructions shackled to an overpowering, blind drive to reproduce.

Maybe the goal of life really is to reproduce, after all. It surely seems that way. Each animal is propelled blindly, unknowingly toward self-perpetuation. They may even do so at great personal sacrifice and risk. We can easily imagine the billions of animals, generation after generation, rushing toward the reproductive prize, all compelled instinctively, genetically to procreate.

But wait, there's a catch here, a hidden agenda, one unrecognized by the individual animal. While the individual organism strives to reproduce, the actual advantage is not for the individual at all but rather for the entire population. The real objective of reproduction *is not self-perpetuation,* but rather the continuation of the species: *species-perpetuation.* You see, it's not the individual that's important genetically, it's the population. It's the population that really matters. Individuals are just the conduits through which the population's immortal genes pass to the next generation. Stated more succinctly, *the individual reproduces so that the species can endure.*

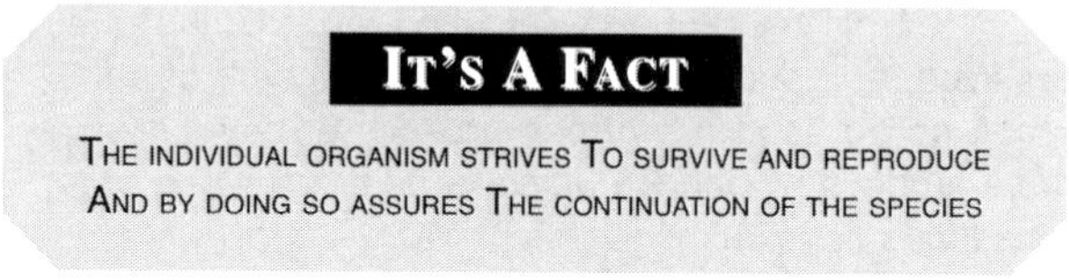

Figure 7.3. The individual strives to survive and to reproduce so that the species can endure.

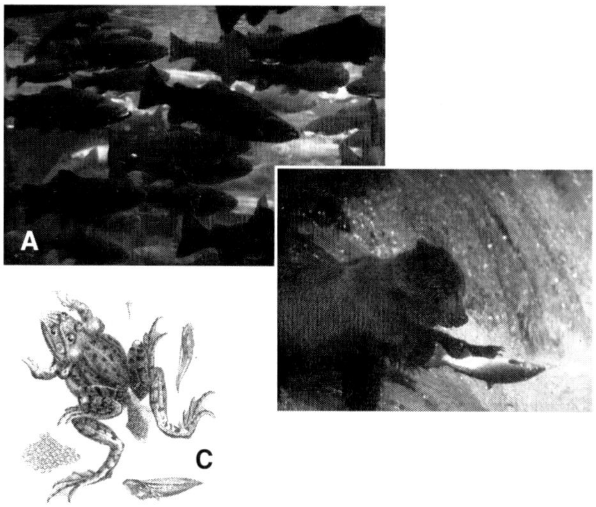

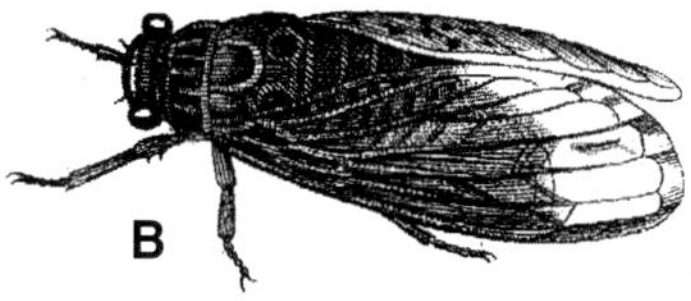

Figure 7.2. Migrating salmon (**A**) gather as a prelude to mass mating after which both males and females die. They must run the gauntlet, including hungry grizzly bears (insert). In **B** above, cicadas emerge from their nymphal case prior to their sessions of mass mating. They, too, all die shortly afterward. A male frog (**C**) grasps a female in a mating embrace called amplexus. This causes her to release an egg mass, which he covers in a cloud of sperm. Of the hundreds of eggs fertilized, only a few will survive long enough to become a reproducing adult.

Topic 8

The Ecological Niche

THE LAW OF COMPETITIVE EXCLUSION STATES THAT EACH SPECIES OCCUPIES ITS OWN, UNIQUE ECOLOGICAL NICHE

There's a concept in biology called the ecological niche that's central to our understanding of biodiversity. In fact, trying to comprehend animal diversity without understanding the niche is like studying genetics with no concept of DNA. The concept has two basic ideas. First of all, *niches are unique for each species of animal.* They are occupied by just one species at a time. Each animal is specifically adapted to exist in its own ecological niche and nowhere else. A successful animal possesses all of the attributes needed to survive and prosper there. It's his private domain. He survives there when no one else can because he is better equipped to compete and will do so at the expense of any intruder. We call this the **law of competitive exclusion**. It's how niches are occupied.

So it sounds then like a niche is a place. Well, it is...but it's much more than just a place. It also includes an animal's essential resources, its physiological and "psychological" requirements. It includes its interactions with other organisms, what type of food it needs, and how it acquires that food. The niche truly includes a complete inventory of an animals' living and nonliving environment. A niche is all inclusive. That's the second principle of the ecological niche—*it includes everything that an animal needs.*

Thus, a popular definition among ecologists is that a niche is the *role an animal plays within its environment.* This definition includes the idea of space, place, behavior, resources, and so on. And, though it's a little vague, it is, nonetheless, very useful in unraveling the mystery of animal diversification.

Niche Adaptation

However, as nice as this definition might be, it's of far greater importance for us to determine, not just what a niche is, but rather how a species becomes customized for its special niche. How do animals become "fit" and why do some become better fit than others? How does a vampire bat become well suited to survive within its niche? How does it acquire the necessary devices to thrive there? The same questions can be asked of snails, and of great horned owls, leeches, tiger salamanders, jumping spiders, armadillos, clams, and human head lice. How does each become so well suited to its respective niche?

Keep in mind that an animal's environment is not static. It, instead, is constantly changing. So, for a species to remain tailored to its niche it too must be able to change, to adjust. We call that give-and-take between a species and its environment **adaptation**, developing a degree of harmony within a changing niche. But adaptation is not like a light switch; it cannot be turned on or off

at a whim. Instead, it is a natural process, often slow, always opportunistic. So, how does it occur? Well, our best calculation so far is that adaptation is the result of three underlying agents: (1) reproduction; (2) competition; and (3) genetic variations.

(1) **Reproduction:** Each animal has received the necessary survival information from its parents through the genetic act of reproduction. Yet reproduction is no haphazard process but is designed in a way to shower the next generation with a constant sprinkling of the best genes available. Thus, through that genetic thread of DNA each succeeding generation receives the **genotypes** necessary to cope with a changing environment. But it doesn't end here. Instead, to succeed, each individual must also compete with its neighbors for survival rights.

(2) **Competition:** To survive, thrive, and reproduce within one's niche, an organism needs to secure a number of essential resources. Among these are food, space, shelter, moisture, and reproductive partners. But resources are not always abundant, and so, to survive, organisms must compete with other members of the same species for their share. Those with the most adaptive attributes, such as size, speed, and alertness, are more successful in this struggle. They can compete more effectively because they are better suited to their ecological niche: they are the ones who stand a better chance to survive; they are the ones most likely to reproduce; they are the ones who will pass on that genetic blueprint to the next generation. And again, what is the nature of that blueprint? It's the very code containing the needed information—some would say, the best information—to survive in one's own niche.

There are, however, limits to what an ecosystem can tolerate. For example, a given area can support only so many organisms...a **carrying capacity** of sorts. Yet animals have an inherent ability to reproduce far beyond what the environment can sustain. Thus, in most cases, there are too many animals competing for too few resources.

Because resources are often scarce, competition for them can become intense. As a result, not all members of a population survive. Some will, others won't. **Differential survival** of this sort is based upon the ability among individuals to compete. Just recall the law of competitive exclusion that declares only one species can occupy a niche at a time. And the result can be stark. Of the millions and millions of eggs produced by an oyster, perhaps only a few, maybe only one or two, will survive themselves to reproduce. The same can be said for most animals: sea anemones, roundworms, ladybugs, geese, alligators, jack rabbits, and so on. Large numbers of eggs or young are produced, but due to the hardships of competition (avoiding accidents, eluding predators, securing food, finding concealment, locating reproductive partners, rearing young, etc.) very few reach adulthood.

Not only does competition serve to control population size but it also allows for the population as a whole to become more vigorous, that is, more fit. It does this by giving some individuals, those with the best attributes, a survival advantage while at the same time culling out those less suited. Successful animals survive to reproduce, unsuccessful ones don't.

Competition therefore favors the most fit within a population and dispenses with the less fit. It serves as a weeding out process to produce smaller yet hardier (fitter) populations. So, even though animal numbers can explode at tremendous rates, competition for rare or limited resources helps to keep populations in check and by doing so also produces individuals even better suited to their niche.

These two factors, differential reproduction and resource competition, play an integral role in niche adaptation. But there's another component that's equally significant. That third factor is **genetic variation**.

(3) **Genes at Work:** We know that animals differ in their genetic makeup. In fact, each sexually reproducing organism, except for identical twins, possesses a different set of genes from every other member of its species. These

Figure 8.1 Competition for food and other limited resources drives adaptive fitness and niche colonization.

personalized sets were received from parents through the biological processes of **meiosis** and **fertilization**. During the meiotic formation of gametes (eggs and sperm), genes are rearranged into different combinations much like shuffling a deck of cards. In this way, every single mating produces genetically different offspring. This reshuffling of genes is termed **genetic recombination** and is a primary source of population variation. Each sexually reproducing animal differs slightly from all others because it possesses a different deck of genes received from two parents.

Although recombination does result in new arrangements of genes, it does not produce any new genetic material. That, instead, is the work of **mutations**. Unlike recombination, which can only reorganize existing genes, mutations actually alter the nature of the DNA molecule. Through heritable mutations, the nucleotide sequence of DNA is modified and that, in turn, causes genotypic changes in individual organisms. These genetic changes themselves may result in small, incremental evolutionary steps within a population—a process known as **microevolution**.

Mutations, by their nature, are spontaneous and incessant. In fact, their frequency can be so high that, at times, they spatter the genetic landscape in a constant barrage. And since they provide the only new source of genetic information, they are vital to the process of niche invasion. They are, in reality, the very raw material upon which populations depend for adaptations—the foundation upon which the pyramid of diversity rests. There is little doubt that mutations are essential to the fitness scheme.

There are, however, a couple of leaks in the mutation bucket. As essential as they are, mutations are not without their shortcomings. First of all, most are **lethal**. The vast majority disrupt life processes to the point that an organism just cannot survive. Since they result in death, these **maladaptive mutations**, of course, quickly disappear from that population. Thus, the great majority of mutations don't contribute one whit to diversity.

The other problem with mutations is that they are **nondirectional**. Just because a niche requires certain adaptations doesn't mean that a mutation will occur to lead a population in that direction. Mutations cannot anticipate the future; they can't know what sort of adaptation might be needed. Nor do they occur in a logical sequence. Instead, they are the spontaneous, random, and rather unpredictable dinner guests at the genetic banquet.

Fortunately, however, some mutations occur that are **nonlethal**. These may prove to be immediately beneficial and may, therefore, thrust an individual to the competitive forefront within a new or changing niche. Indeed, these favorable mutations provide the building blocks upon which adaptation and niche invasion depends. A beneficial mutation within a population may improve the ability of some individuals to compete and thus, step-by-step, the population becomes increasingly better adapted to its niche. On the other hand, nonlethal mutations may be neutral and, as such, just linger within the population, doing no good, but also doing no harm. These **lingering mutations** may provide a genetic reservoir by which members of a population can later take advantage of a new niche and its myriad requirements.

IT'S A FACT

MUTATIONS ARE SPONTANEOUS, PERMANENT CHANGES IN THE GENETIC CODE. THESE ALTERATIONS IN THE BASE SEQUENCES OF THE **DNA** MOLECULE PROVIDE THE RAW MATERIAL, THE FUEL UPON WHICH THE FIRE OF FITNESS BURNS.

THE SURVIVAL FORMULA

1. RECEIVE GENETIC INSTRUCTIONS FROM YOUR PARENTS.
2. COMPETE FOR NECESSARY RESOURCES.
3. THE MOST FIT SURVIVE AND REPRODUCE.
4. POPULATIONS ACCUMULATE ADAPTIVE MUTATIONS.
5. SURVIVAL GENES ARE TRANSMITTED TO THE NEXT GENERATION.

The Diversity Connection

What does all of this have to do with animal diversity? Well, everything. Diversity is nothing more than the net result of animals striving to fill ecological niches. The more niches, the greater the diversity. If there are indeed 10 million niches, then there can be 10 million different species of animals, each striving to occupy one for itself.

New niches emerge continually through a variety of mechanisms: extinctions, geological forces, climactic instability, diseases, starvation, human interference, and so on. When niches become available, animals seek to occupy them. *Nature abhors a void* is an old expression that communicates this effort. But how does that happen? The best explanation is that it occurs through a process known as **preadaptation**. Animals who already possess the skills, techniques, and structures needed in the new niche will be the ones to colonize and survive there. Those lacking the necessary skills cannot.

Figure 8.2 Certain individuals have an advantage over others in the competition for limited resources. They tend to have a higher rate of survival. Their genes will then be passed to the next generation.

The next logical question then is "Just how does an animal become preadapted?" Here, again the answer is quite clear...through mutations. This is how it works. Nonlethal mutations occur constantly within populations. Some of these mutations are not immediately beneficial in the present niche but, since they are nonlethal, they may linger within the population at large. If these lingering (neutral) mutations provide an organism the slightest advantage in a new situation then, it is already partially adapted to survive there. It is ready to colonize a new niche even before the niche materializes. And this invasion can happen rather swiftly. Why? For the simple reason that the organism is, to some degree, already preadapted to survive there.

However, mutations, as we have also stated, are random and nondirectional. So, in this respect, mutations, unlike reproduction, are very much a product of chance. Does this mean that adaptation and niche fitness relies on genetic randomness? To a great extent, yes. But in case you think trial and error cannot provide the impetus for niche invasion, you need only reflect upon the enormous numbers of individuals we are dealing with, the relatively high rate of mutations, and the millions of years over which these events occur.

Summary:

Even though mutations are random and nondirectional, they are the only source for new genetic information. They, therefore, provide the raw material for natural selection to occur. In other words, mutations lead to preadaptation.

Where Did They All Come From?

"The Noiseless Foot of Time"

Can you just imagine, 10 million different kinds of animals, maybe more? Where did they all come from and how did they get here? Why these particular 10 million and not some other combination? What makes one type prevail and another fail? How long did it take to reach this level of diversity? Where do we go from here? It seems that our questions are endless. So where can we look for the answers? Well, one good place to go when you want answers is to a depository of information... a library. But is there any such reservoir of animal records? Fortunately, there is. The history of some animals has been well recorded, preserved in a stone library. We know it as the fossil record. Throughout time, because of a combination of factors (body design, habitat, internal or external skeletons, etc.), certain organisms have been prone to fossilization. Their past is documented in this rocky record. They have just been waiting through the eons for us to decipher their cryptic messages. As we pick, scrape, brush, and reconstruct the past, we are slowly reading that ancient book. And what we have uncovered is basic to our understanding of animal diversity. The record tells us what was present and what was absent, and if read accurately, it gives us a glimpse into the very lives of those ancient animals and, more importantly, it gives us insight into the very process of species formation. Unlike the Shakespearean quote above, the silent record is really not so noiseless after all; on the contrary, theirs is a small monstrous voice!

Speciation

New species originate along two pathways, and like a junction in a road, these two paths may lead in totally different directions. We find that one route, under proper conditions, can produce a profusion of new species, while the opposite path, produces but one. Similar to a tree, the first path branches out into new niches in a radiating pattern. Here, one ancestral parent species is able to produce a plethora of new genotypes to invade a variety of niches. Each branch of this ancestral tree may ultimately result in the formation of a new species. In other words, many new forms can spring from a common stem, like twigs on the same bush. This form of speciation is called **adaptive radiation** and, much like smoke emanating from a fire, it permeates its surrounding. At times, this radiation may produce only a handful of new species then it stops. But at other times, the process may literally explode into a veritable landslide of new forms. Such explosive radiations are common, for instance, on remote islands as well as in the aftermath of drastic environmental or geologic changes.

The second path to speciation occurs when one population responds to environmental changes and, over time, is transformed into a single new species. Most often in this case of **linear descent**, the original species disappears from the scene leaving only a lone offshoot. Note that, unlike adaptive radiation, linear descent produces no net increase in the number of animals, just the appearance of a single new species.

In both cases, however, speciation results from the same force: accumulated mutations, which, in turn, leads to the appearance of new morphological types or **phenotypes**. These phenotypes may, in time, lead to

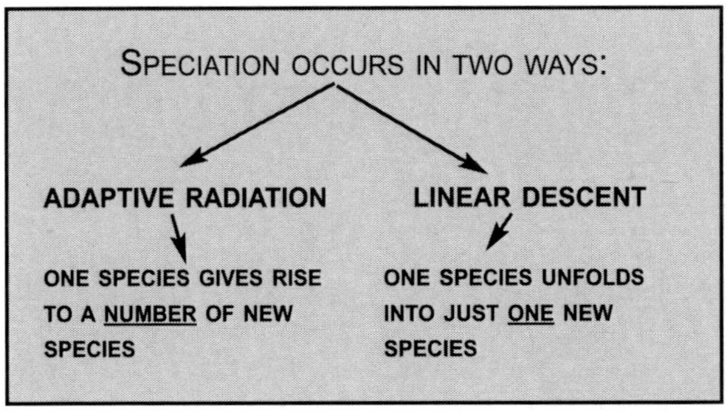

more pronounced changes—a process termed **macroevolution**. These cumulative changes may produce such distinctive phenotypes that eventually an entirely new biological species appears. But the process can be agonizingly slow. In fact, it may take millions of years of accumulated adaptations for a single new species to emerge in this way. Nonetheless, scientists believe that this is exactly how many, if not most, of our animals originated—through a slow-paced mechanism called **gradualism**.

However, gradualism is not the only way that new species may arise. In fact, recent interpretation of the fossil record supports another possibility, **punctuated equilibrium**. There are apparently circumstances in which macroevolution occurs rapidly, in spurts, resulting in an avalanche of new species. These radiations often follow massive extinction episodes when a profusion of new niches become available. Certain preadapted members of the population can quickly spread into these newly formed niches. These radiating spurts are then followed by relatively long periods of evolutionary inactivity or stasis, followed by yet another spurt, and so on. An unsteady, punctuated equilibrium may then be achieved.

Natural Selection

most impt. to fill open/new niches.

Overview

The appearance of new species occurs through two mechanisms: the usual way seems to be through a slow, lingering, step-by-step process called **gradualism.** However, it also seems that new species may literally erupt in the blink of the geologic eye. These spurts are followed by periods of relative evolutionary calm; a **punctuated equilibrium** dominates. But the actual process is always through ***natural selection.***

When a new niche appears, you can bet that an animal will follow to occupy it. New niches surface constantly and animals respond by relentlessly filling them. It takes a number of steps for this to occur: reproduction plays a big part; competition is essential; and mutations are indispensable. But there is one more step involved, **natural selection**. This is a simple but absolutely crucial step. However, before we discuss it, let's first review what we have learned so far.

(1) We know that mutations are the source of genetic change. They occur spontaneously, randomly, and continually. Although they have no predictive capability, they, nonetheless, provide the very means of adapting to changing niches. We have further learned that there is no known way to produce a particular mutation. No organism, for example, can command a given mutation to produce a desired outcome. No animal can simply push a button and cause a needed genetic change.

Thus, speciation may be slow or rapid, subtle or explosive. Which one predominates in a given situation depends on both internal (mutation rates) and external (extinction rates) forces. When niches become available on a massive scale due, for example, to a catastrophic climactic event, then speciation appears to speed up. Niches are then colonized rather quickly. On the other hand, under more stable environmental conditions, the process is much more gradual and relies on slow, accumulating microevolution.

(2) We have also learned that for mutations to have a genetic influence they must occur within the reproductive tissue, the testes or ovaries. Regardless of what happens to a structure (such as an appendage, antenna, liver, or eyeball) during an individual's lifetime, that change cannot be transmitted to the next generation. Only if a mutation occurred within the reproductive (gonadal) tissue will changes be inherited. Thus, the one-time popular concept of Lamarkian "use and disuse" has been discredited as a player in evolution's creative stage.

(3) We have learned that the only route open to invade a new niche is through preadaptation. The inheritable advantage must be present in the population before invasion of the new niche can occur.

(4) Finally, competition for limited resources is a fundamental step in evolution. An animal must survive long enough to reproduce, long enough to pass on the survival traits or mutated genes. Otherwise, adaptation cannot and does not occur. An animal does this by out-competing its neighbors for essential resources. If successful, it will reproduce and pass the successful features to its offspring.

Summary

1. Mutations are the source of all genetic change
2. Mutations must occur in reproductive tissue
3. Preadaptation is the key to the evolutionary process
4. Animals compete for limited resources which leads to natural selection

So again, we come to the all-important question, "How does an animal get the necessary features to invade a new ecological niche?" Well, the only possibility is that it must rely upon genetic transformation, recombinations, and random mutations that have already occurred. In other words, to invade a niche an animal must already possess the needed attributes. It must be preadapted for survival in that niche. Then as more preadapted individuals flood into these niches, or as more mutations occur, the population ultimately acquires the necessary skills, structures, and abilities for that particular niche. They accumulate a competitive repertoire of sorts. As a consequence, they are the ones that have a higher probability of surviving and reproducing. Their traits are selected, through differential reproduction, to be passed on. This is the final step in our diversity equation: selection of fit traits, or in other words, **natural selection**. Fit individuals survive (are selected) and pass on their skills to the next generation. The unfit don't.

Natural selection, then, is the ultimate referee in evolution's genetic contest. It serves as a "judge" by which the fit pass muster and the unfit don't. But natural selection is a two-way street. It rewards the fit at the expense of the unfit. In fact, many biologists refer to natural selection as **nonrandom elimination**. By this they are observing that not only does natural selection choose who survives, but it also chooses who doesn't.

It's important to note here that in natural selection there is a premium placed on those characteristics that lead to fitness, and those sets of characteristics are different for each ecological niche. Note also that, while mutations occur randomly within individual organisms, natural selection itself is *nonrandom*. Chance may be instrumental in the appearance of mutations, but natural selection works through anti-chance; that is, it favors the preadapted individual and thus is not random at all. If natural selection occurred through luck or chance then each individual in a population would have an equal opportunity to pass on its genes. However, we know all too well that is not the case. On the contrary, it's the fit, not the lucky, that tend to reproduce. Natural selection chooses the best set of traits for a given niche; it fine-tunes every population to fit better and better within its particular niche.

One final and important point about evolution. *It's not the individual but the population that evolves.* Individuals sow mutations but the population reaps the evolutionary harvest. From the vast variations within a population, natural selection chooses only those with the means to survive. Whether slow or fast, plodding or explosive, natural selection does, indeed, create new species of animals. And so, through the combined processes of mutations, genetic recombination, preadaptation, competition, and natural selection, biodiversity continues its universal unfolding.

Figure 9.1 At times, natural selection seems to prefer the bizarre or exotic. However, while some features may seem strange to us, keep in mind that whatever offers survival advantage in a given niche will most likely be selected.

How Do We Know?

All of this sounds fascinating, a rather logical explanation of the biodiversity question. But do we really know that biodiversity originates in the way we have described? Are mutations, genetic recombination, competition, and natural selection the true agents of diversity? The answer is that we are reasonably sure they are, but not absolutely sure. We can surmise and deduce, we can use our educated rationale, and we can make scientific assumptions, but we cannot know absolutely. No one was there during those millions of years to witness and record the emergence of different lifeforms. No one can predict just how, when, or if certain mutations might occur. No one can even measure the degree of fitness required for any particular niche. So, while evidence does exist to support our model, we cannot know with certainty.

What evidence? Well, there is actually an abundance of proof in favor of this process. But because the process takes so long, the evidence is, for the most part, indirect. Indirect...but powerful. So powerful, in fact, that few serious biologists doubt its veracity. This evidence comes to us through a number of separate, yet interconnected routes: the geological record, fossils, the geographical distribution of animals, comparative anatomy, comparative embryology, artificial selection, and direct observation. Let's examine several of these.

The Fossil Record

We know from literally thousands of discovered and radioactive dated fossils that many life-forms no longer exist. We know that certain animals are prominent or even abundant in fossil form but have completely disappeared as living entities. Just think of some of the more famous extinctions: dinosaurs, archaeopteryx, trilobites, passenger pigeons, hairy mammoths, giant armadillos, saber-toothed tigers, giant sloths, the Irish elk, and an endless stream of others. All are gone. Fossils alone bear their ancient footprints.

In contrast, representatives of the great majority of living animals (those on earth right now) are absent from the fossil record. They apparently did not exist in those prehistoric days. They have evolved since then; they are newcomers to the world's fauna. So we can reasonably conclude that biodiversity is vastly different today from what it was in ancient times. The animal spectrum has, indeed, changed. About that there is little doubt.

Not only are there different kinds of animals alive now, but they occur in a much greater variety than ever before. By comparing living animal populations to the fossil record, it is apparent that diversity is considerably greater now than at any time in the past. New animals continue to appear in radiating fashion. Like lava flowing from a volcano, natural selection has produced an ever-expanding stream of animals, resulting in the 10 million or so species thought to exist today. By listening to their petrified echos, we can learn much of history that fossils are longing to tell.

Comparative Anatomy

The essence of diversity is that every species differs from every other one. And while that is correct, the opposite is also true: animals are also very much alike. For example, they all use DNA as the purveyor of genetic information; they all share the same 20 amino acids to construct the magical protein molecules; they all contain cells with a similar molecular makeup; they all require the same biochemical materials (e.g., calcium, phosphate, sodium, potassium). All animals are also heterotrophic, in that they must obtain their energy from external, rather than internal, sources.

Furthermore, many animals have similar body designs. The forelimbs of amphibians, reptiles, birds, and mammals show structural similarities. Even limbs that

Trilobite (extinct)

Sabertooth Cat (extinct)

Figure 9.2 The fossil record has shown that both the number and kinds of animals have changed drastically throughout time. Most organisms that once existed are gone; others that didn't exist in the past now flourish.

appear much different from each other are based on a common pattern. For example, legs, wings, and flippers all have a basic anatomical construction.

Many other body parts show correlations: location and action of various muscles; uniformity within digestive systems; similarities in nervous systems and photoreceptors (eyes); even body coverings show interconnections...scales, feathers, and fur have common roots, for example. The list of structural similarities is a very, very long one.

These morphological associations, in fact, form the basis for the important concept of **common ancestry**. When two or more animals show common designs, engineering, or features, we begin to wonder if they have descended from a common source. There is the real likelihood that we are seeing the result of **adaptive radiation**. The closer the resemblance, the closer the possible evolutionary relationship.

Then there's the question of **vestigial** structures. These are features that have lost their function, or their function is drastically reduced, yet they linger in some animals. Are they remnants of ancient usage giving us some insight into what an animal was like in the past? Maybe. Consider the following structures in the human: (1) the *lacrimal caruncle* (the little reddish mass in the corner of your eye considered by some to be a remnant of the nictitating membrane, or third eyelid, of reptiles and birds); (2) muscles in some people that move the external ear; (3) the *coccyx* or tailbone; and (4) the *arrector pili* muscles that cause our body hairs to "stand on end" (erect hairs are important in other mammals, making an individual appear larger and more ferocious and even trapping air for insulation, but they have no known function in humans). All of these structures have either lost their original function or their function is greatly reduced in adult humans, yet each persists.

Vestigial structures are evident in other animals as well. For instance, in some snakes and whales, the internal remnants of a pelvic girdle remain, even though they have long ago lost their hind limbs. Many birds have become flightless, yet reduced, stubby wings remain. Also, cave-dwelling fish and subterranean moles retain their residual, sightless eyes. Perhaps these structures, too, reflect an evolutionary link to the past. Maybe they really do open a window and let us peer, furtively, into our yesterday.

Common Ancestry

Animals have an ancestral connection—they are more alike than they are different. For example, animals are united with each other through their molecular makeup, their cellular machinery, and their genetic thread. They also have similar metabolic and nutritive needs; their reproductive patterns are similar. Many have comparable body design. Some claim a closer affinity with each other, while others are more distantly related. Yet all have a shared ancestry woven within their protoplasmic fabric.

Animals are truly different...but related. They share a common lineage. Each animal alive today is connected to its predecessors; each is linked, in one way or another, to that endless, ancient evolutionary chain.

Comparative Embryology

Embryology is the study of an organism's early growth and development. It traces an individual's progress from conception through each formative stage, to the time of hatching or birth. Through these detailed events we have discovered the wonders of development: gamete formation, fertilization, gene control, tissue formation, organ formation, and birth.

When a variety of embryos are compared, it is clear that there are vast developmental similarities. For example, the actual sequence of development is almost universal. It involves a number of stages including fertilization, cleavage, blastulation, gastrulation, histiogenesis, and organogenesis. These are but a few of the rather technical terms used to describe an animal's development, and the pattern is uniform throughout most animal groups. In other words, the developmental forces seem to be the same for each animal. The same evolutionary mechanisms seem to be at work in the embryo as well as in the adult.

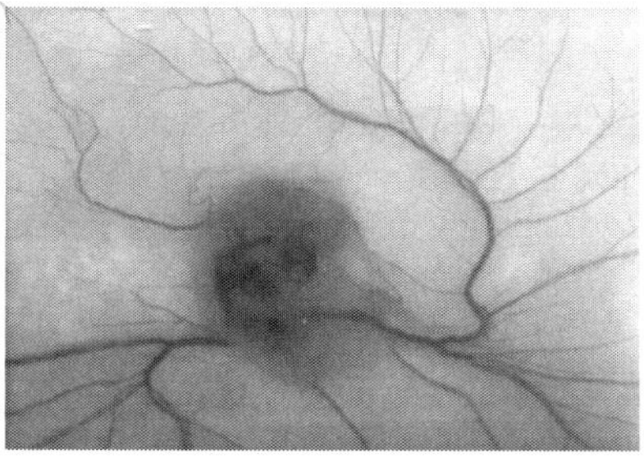

MANY EMBRYOS OF HIGHER ANIMALS PASS THROUGH SIMILAR STAGES OF DEVELOPMENT WHICH TENDS TO SUPPORT THE IDEA OF A COMMON ANCESTRY

Geographical Distribution

Of all the evidence in favor of natural selection, the distribution of animals upon our planet is perhaps the most compelling. A study of zoogeography unveils some rather surprising findings.

First of all, throughout the world, regardless of location, similar habitats are occupied by physically similar animals. Take, for instance, the North American prairie. Animals suited for survival there, such as the pronghorn antelope, the coyote, or the prairie dog, are similar in structure, size, and behavior to animals in prairie ecosystems

Figure 9.3 Even though the Asian male Mandarin duck (top) and the North American wood duck live on different continents, they occupy similar habitats and show remarkably similar body designs and coloration.

throughout South America and the African continent. The same is true for animals living in deciduous forests, ponds, lakes, swamps, tropical rainforests, mountains, and deserts. Most have their ecological counterparts in other areas of the world. The Alaskan brown bear is similar to the Russian brown bear; the North American golden eagle is similar to the white-tailed eagle of northern Europe and Asia; the Tasmanian wolf is similar to the North American timber wolf; the South American condor is similar to the California condor; New World and Old World monkeys resemble each other; and so on and so on.

When two habitats are alike, their occupants, too, often share a surprising similarity. In fact, the uniformity is sometimes astonishing. Mutations, competition, reproduction, and natural selection have created "look-alike" animals in similar ecosystems throughout the world. This is called **convergent evolution** in which similar habitats produce similar life-forms. Examples are endless, the correlations irrefutable, the conclusion inescapable. The same evolutionary forces are at work regardless of locale.

But let's not get carried away! For it's just as important to recognize that, while occupants of similar habitats are themselves similar, they are not the same. In fact, they differ in virtually every case. An animal observed in a Madagascar rain forest will be comparable to, but different from, an animal occupying a corresponding ecosystem in Central America. Why is that? One would think that if the habitats were the same and if evolutionary process are, indeed, universal, then the animals occupying them would, likewise, be the same.

To think that way, however, is to discount the fact that although habitats seem to be identical, the ecological niches are not; each is unique. Also, evolution, as we have seen, is driven very much by random, chance events. Thus, you would not expect any two regions, regardless of how similar their niches might be, to produce identical species. Each niche requires a different set of adaptations. Habitats such as prairies, rain forests, and deserts produce similar, but unique, lifeforms because each habitat contains unique ecological niches. For example, horns may be an advantage to animals living in comparable ecosystems, say a prairie. Thus, prairie-living antelopes develop horns. However, mutations and natural selection will dictate the exact type of horn in each instance. The outcome will be similar, but different, as in the figure below.

I think there are two important considerations here. First, the forces are the same for each niche: mutations, competition, survival, and reproduction work together to produce similar, yet unique animals. Secondly we learn that natural selection favors success, and success can mean different things in different niches. Thus, while animals in parallel niches may be very much alike, they are nonetheless, also very different. Evolution does not have an "ideal" size, shape, or body plan. It just selects those best fit in a given situation.

Figure 9.4 Animals occupying similar habitats, even continents apart, show comparable body design. Here pronghorn (*Antilocarpa americana*) from the North American prairie (left) is contrasted with springbok (*Antidorcas marsupialis*) of south Africa. Note that they are similar but different.

IT'S A FACT

THE GOAL OF EVOLUTION IS NOT PERFECTION...IT IS SIMPLY TO SELECT THE MOST FIT INDIVIDUALS FROM AMONG THOSE COMPETING FOR ANY GIVEN RESOURCE.

Island Biogeography

The distribution of animals on islands provides additional insight into the evolutionary process. In fact, this feature probably had as much impact on Darwin as any other as he pieced together his now famous theory.

Some conspicuous and interesting things happen to animals when they invade island ecosystems. Because competition, predation, and diseases are often absent or reduced on islands, a colonizing life-form can experience pronounced radiations in a rather short period of time. Empty niches are often plentiful on islands and relatively easy to invade. In fact, a single colonizing animal can disperse into a variety of available niches like ripples on a glassy pool.

That was the case on the Galapagos Islands visited by Darwin in the mid-1800s. On this small archipelago off the coast of Argentina he discovered several episodes

Figure 9.5 The Galapagos Islands played an important role in our understanding of the evolutionary process.

of island radiation. For example, he observed several species of birds that had apparently sprung from a small founding group of finches. Throughout the world finches are small, seed-eating birds. They have stout, conical beaks engineered to crush hard seed coats. Darwin found these basic, seed-eating finches on the islands. But he also recognized a number of finch-like birds that had strange, new beak types: a chisel-like woodpecker beak; a second one now razor-sharp and designed to feed, vampire like, on blood; an elongated beak meant to probe into small crevices; yet another one intended to catch insects; and another one devised to feed on flower nectar. He discovered thirteen different designs in all—thirteen different species of finch-like birds derived from a single finch-like ancestor. He saw this genetic divergence as powerful evidence for his developing theory of natural selection.

Darwin found similar radiations of birds and other animals on each island visited in the chain. It seems that each island had gone through its own version of evolution. This, he realized, was the final piece of the evolution puzzle. These islands were fabulous, living laboratories where natural selection had created, on each, a unique and diverse animal community.

Artificial Selection

Further confirmation of evolution comes from human manipulation of animal genetics. Through the centuries man has domesticated a number of animals and plants for his own use: for food, sport, labor, clothing, companionship, flower gardens, and now, it appears, even for organ implants. Breeders and scientists alike are constantly discovering and manipulating the phenotypes of domestic plants and animals. By doing so they can select the very traits deemed desirable in each species: a full slab of ribs, a swift hawk for falconry, sweet cow's milk, a luxurious fur stole, a work horse, a hairless cat, a miniature poodle, a floppy-eared rabbit, a new variety of rose or tulip, a talking parrot, beasts of burden, and even pigs for heart and liver transplants!

This artificial selection in the laboratory, garden or barnyard mirrors natural selection in nature. The operating forces are the same: mutations occur; some resulting traits are selected by a breeder or a farmer, others are neglected; animals with the desired features are permitted to breed while the undesirables are not. As the coveted traits eventually spread throughout the population, those preferred traits in each animal become the norm.

Figure 9.6 Relying on the principles of organic evolution, humans have “invented” an amazing variety of domesticated plants and animals. From floppy-eared rabbits to wooly sheep, artificial selection has underscored the processes of natural selection.

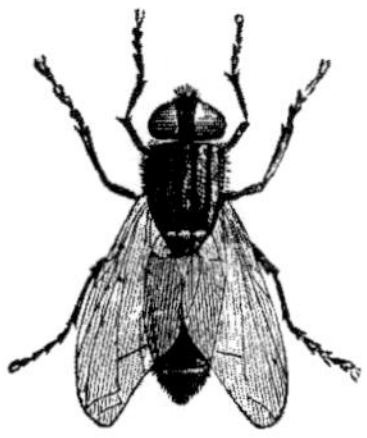

Direct Observation

One of the earliest and most basic questions asked about the evolutionary theory was, "Is it happening right now, and if so can we see it?" For almost 100 years the answer has been, "Well, yes...and no." But that has changed. Over the last 50 years or so we have been keeping a scorecard, recording each instance of possible evolution. The results are now coming in, and they show, clearly, that we have, indeed, witnessed evolution in progress.

Take houseflies for instance...those ubiquitous nuisances. They have been our constant and detested companions for thousands of years. And for just as many years we have tried everything imaginable to wipe them off the face of the earth. By the middle of the 20th century the best we could come up with was the flyswatter and sticky flypaper. Then along came the insecticide DDT. We had, at long last, discovered a remedy for this torment. Housewives everywhere, armed with their "bug bombs," staged an onslaught. Billions upon billions upon billions of unsuspecting flies fell victim to DDT's lethal mist. We had finally conquered this squalid, disease-carrying, maggot-producing, filth-spreading pest. But, had we?

Unknown to us, the lowly housefly had a trick up one of its six sleeves. The problem was that we didn't strike a fatal blow when we had the chance. In our assault, we didn't know that some flies were less affected by DDT than others and they had survived the purge. *They survived because some were naturally resistant to DDT's poison.* They were preadapted. These survived in small pockets, here and there, and that resistance was gradually transmitted from one generation, then on to the next, and so on. The rest is history. By the early 1970s houseflies were virtually unaffected by DDT exposure. Flies survived, DDT did not. Flies had evolved while DDT is gone.

We have seen many, many similar examples of evolution in progress. Since insects reproduce so rapidly and expansively, they have been the most obvious targets of our observations. But we have seen it elsewhere, in bacteria, plants, and even viruses (the AIDS virus, for example). It seems that evolution, especially microevolution, occurs constantly and is sometimes observable.

Evidence for Evolution: A Summary

Evidence for the natural selection paradigm of evolution comes from a number of quarters. Collectively, they make a compelling case in favor of this model:

(1) **Fossils:** The fossil record shows that the kinds and numbers of animals have changed throughout time. It also reveals a number of mass extinctions followed by progressive increases in diversity.

(2) **Comparative Anatomy:** Animal body design, molecular makeup, and biological machinery show such similarities that common ancestry is uniformly accepted among zoologists.

(3) **Comparative Embryology:** Developmental patterns among plants and animals indicate strong interconnections. Also, many embryos show features seemingly reflective of their evolutionary past, a concept called biogenesis.

(4) **Animal Distribution:** The geographic distribution of animals on earth provides strong evidence that the principles of natural selection are universal: similar habitats produce similar organisms. On the other hand, animals occupying similar habitats are alike, but different. This indicates that each ecological niche is unique and, in turn, produces unique animals. Island ecosystems are conspicuous in their display of these features.

(5) **Artificial Selection:** The domestic production of plants and animals parallel evolution in that humans select those traits deemed desirable, just as natural selection works in nature on fit individuals.

(6) **Direct Observation:** An increasing body of data derived from direct observation (pesticide-resistant insects, resistant pathogens) have further bolstered the natural selection model.

Now How Does It Work, Again?

The theory of natural selection, as proposed by Charles Darwin and later modified by many others, can be rather confusing at first. After all, it is a concept meant to explain the very mechanism of biodiversity on our planet. It is trying to answer the age-old question: How have all organisms, in their endless variety, come to be?

You could expect, therefore, that such an ambitious, comprehensive, and all-encompassing theory would be complex and difficult to grasp. But that is not the case at all. Indeed, what is surprising about the evolutionary concept is not its complexity, but rather, its fundamental simplicity. The theory of evolution can be summarized in just six, easy-to-understand principles. It's supported by six biotic pillars.

The Six Pillars of Evolution

(1) Individuals within a given species have *inherited differences;* they vary from each other in subtle, but important, ways.

(2) Plants and animals have a *great reproductive potential,* often beyond that which a given habitat can sustain.

(3) But *natural resources* (food, space, etc.) *are limited.*

(4) Since population numbers can outstrip a habitat's carrying capacity, organisms must *compete* for these limited environmental resources.

(5) *Those individuals better suited* to their niche, because of inherited beneficial traits, compete more effectively, and *stand a greater probability of surviving and reproducing;* they will be the ones to see their genes pass through the filter of natural selection.

(6) Because of competition and natural selection, population size over the long term remains relatively stable, fitness is improved (due to differential survival), phenotypes change (due to differential reproduction and accumulated mutations), and eventually *new species are launched.*

Topic 10

The Specialization Paradox

As we have learned, the more specialized a group becomes the better it fits into its niche. So there's an advantage for a species to become highly specialized, right? Well, maybe not. Ironically, there's a potential, but ever so real, danger that if a population goes too far in its race toward specialization it may be drawn into a trap, the specialization paradox. Here's what might happen. On the one hand, the pressure is to become more and more specialized because survival rates increase for those groups that do so. On the other hand, if the specialization is along very narrow lines the group may lose its ability to withstand even a minor environmental change. Indeed, it can easily become entangled in a web of no return. If that happens extinction is sure to result.

Parasites often face such a dilemma. They tend to become restricted to a single host species and will only be found on that one host. If that happens, the parasite's fate becomes irreversibly linked to that of its host. If the host numbers decline, so do the numbers of the parasite. If the host vanishes altogether, so does the parasite. Overspecialization is a clear problem for many parasites as it is for a multitude of other animals.

A comparison between two similar species, the Giant panda *(Ailuropoda melanoleuca)* and the American black bear *(Ursus americanus)* further illustrates this dilemma. The panda feeds almost exclusively on bamboo, which quickly passes through its digestive system so that much of the nutrients are lost. Thus, the panda must eat large amounts each day. The problem is that bamboo supplies are becoming scarce as people replace them with agricultural crops, and due to natural bamboo cycles. As a result, the panda, because it's so specialized, faces a bleak future (fewer than 1,000 exist in the wild).

On the other hand, the black bear is much less selective in its feeding habits. As an omnivore it will eat virtually anything from insects, to vegetation, to small game, to human garbage. Thus, it is a much more adaptable animal and the population of black bears is actually on the rise, even in the midst of tremendous human encroachment.

Figure 10.1. The panda and the black bear show the extremes of the specialization paradox. Pandas are specialized to feed only on bamboo whereas black bears accept a wide range of food. As a result, the panda populations are declining as bamboo supplies dwindle but the black bear continues to thrive even in the face of considerable human intrusion.

So the trick is to be specialized, but specialized to a broad range of conditions rather than to just one or two. For instance, a parasite might specialize on birds in general rather than on one species of bird. However, there's danger lurking here, too. For if a population becomes overly generalized, it lessens its ability to compete... that's the paradox. It must be specialized but not overspecialized; it must be generalized but not overgeneralized. If the species cannot solve that puzzle, it is destined to travel along that one-way path of no return.

Solving the Specialization Puzzle

When we consider the long-term survivors, the "living fossils," such as the centipede, the dragonfly, scorpions, alligators, opossums, and sharks, we find they have all been able to solve the mystery of the specialization paradox. They have balanced the equation well, being specialists and generalists at the same time. Through the eons they have withstood the countless pitfalls, hazards, and catastrophes of the ecological landscape. They have survived because of their genetic elasticity. They have persevered and have left the overspecialists and the overgeneralists behind, encased in their fossilized tombs.

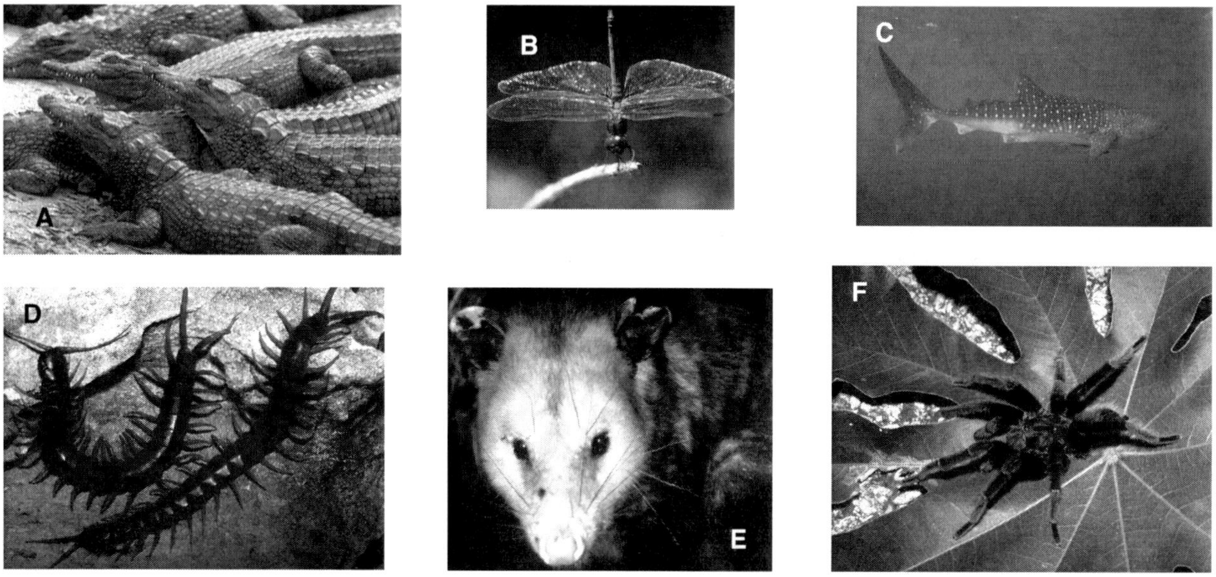

Figure 10.2. Many animals have solved the specialization paradox by balancing their degree of generalization against their level of specialization. Crocodiles (**A**), dragonflies (**B**), sharks (**C**), centipedes (**D**), opossum (**E**), and tarantulas (**F**) all have survived the millennia by solving this paradox.

Topic 11

Form Follows Function

INDIVIDUALS SOW MUTATIONS, BUT IT'S THE POPULATION THAT REAPS THE EVOLUTIONARY HARVEST

Let's see how this process of natural selection might work through a simple, hypothetical example. Suppose that in a large population of mammals some individuals undergo a specific, heritable mutation. The result of this genetic alteration is that these animals inherit a slightly thicker coat of fur. Now suppose that the environment, too, changes slightly: it becomes just a little cooler. It's not unreasonable to assume that those individuals with thicker fur will now have a competitive advantage. As a result, they have a greater probability of surviving and reproducing than their less furry comrades. They will then pass on their "fur" genes to their young, who will do likewise. In time, assuming the climate doesn't change again, those "fur" genes will be broadcast throughout the entire population. Eventually, all members will become furrier.

Now let's suppose that a new player arrives on the scene, a vicious predator. A thick coat of fur might help to some degree but not nearly as much as, say, a coat of armor or even spines. If a mutation had already occurred in the population in which hair was slightly modified into a scale or a spinelike covering, then those animals would now have the advantage and they would more likely survive to reproduce.

We could imagine similar scenarios for an endless number of features: say, an elongated tail, or maybe larger canine teeth, or perhaps larger eyes, louder vocalization, longer legs, improved eyesight, maybe an enlarged cerebral cortex...and on and on, until at some point the mammal we began with no longer exists, but is now something quite different, *a new species.*

But regardless of the outcome, remember that the forces are always the same: heritable mutations, which lead to enhanced survival and reproduction.

Note that the process followed this course:

(1) A specific, inheritable mutation occurred

(2) The environment changed, or a new predator arrived

(3) Those with the mutations had an advantage; they were preadapted to the new situations.

(4) They survived and reproduced at a higher rate than others

(5) The "furrier," "spinier," or "scalier" phenotype was selected and eventually predominated.

Also note that:

√ The mutation was random; it did not anticipate the change in climate or the arrival of a new predator.

√ The mutation must occur in reproductive tissue to become part of the genetic program.

√ Those individuals without the mutation were at a competitive disadvantage.

√ Natural selection works by filtering out the fit from the unfit within a particular niche.

Topic 12

The March to Land

Introduction

There's few who would argue that of all the animals on earth arthropods are the most successful. Regardless of how success might be measured, arthropods prevail. If diversity is the yardstick, arthropods win; 8 out of every 10 known animals is an arthropod. If habitat usage is the barometer, arthropods dominate; they have invaded virtually every known environment and monopolize an incomprehensible number of ecological niches. If sheer numbers is the standard, arthropods win again. They are among the most prolific of animals. They "spread like flies." In fact, using the housefly as an example, it seems that more of them hatch each day than there are humans on earth. Or consider the termite or the ant. It is estimated that there are 1,500 pounds of termites and one million ants for every person alive! A single termite queen lays 30,000 eggs per day (that's one every three seconds). She lies in her underground boudoir, fed, cleansed, pampered, and all the while incessantly pumping out eggs. And what about locusts? They emerge in sun-eclipsing clouds of over 50 thousand million individuals, eating and reproducing their way to ignominy. One swarm can consume over 3,000 tons of vegetation per day!

Arthropods have, indeed, discovered what is needed to survive, and they do it better than anything else. I've studied insects for a long time and I'm amazed at their versatility, resourcefulness, and resilience. No matter what we do to control or eliminate them, they rebound as tough as ever, ready, it seems, to tackle any assault we might deliver. In the U.S. alone, we spend $250 million each year to unsuccessfully annihilate cockroaches. They continue to thrive while we, on the other hand, seem intent on destroying the very natural bounty on which we depend. We despoil natural habitats at an appalling rate. Eliminating, all the while, the very diversity that gives earth its vitality. And who will survive the onslaught? Why, arthropods, of course.

This recurring idea of human extinction followed by the takeover by swarming hordes of arthropods was graphically expressed years ago:

it won't be long now, it won't be long
man is making deserts of the earth
it won't be long now
before man will have it used up
so that nothing but ants
and centipedes and scorpions
can find a living on it.

-Marquis

We've all heard about an impending revolt by spiders, scorpions, and insects. But is it real, or just science fiction hype? Is it possible that arthropods will take over, and if so, when will it happen? Finding the answer to that question is much like untying the Gordian knot; it's hard to know where to begin and you're a little afraid of the outcome once the question is unraveled. However, according to famed biologist, E.O. Wilson, who has studied this question perhaps more than anyone, it's not a matter of when arthropods will inherit the wind for, in his words, "they already have!"

Figure 12.1 Ants are among the most successful of all animals. They exist in huge, social colonies that number well into the millions of individuals.

Changing Earth's Tapestry

Many millions of years ago animals accomplished a feat so momentous that it changed the very nature of our planet: they invaded land. And it didn't happen just once, but twice. Much has been made of amphibians as the source of land animals; of how they evolved from lungfish and eventually gave rise to the reptiles and then onward to birds and mammals. We marvel at their skills, their initiative, their boldness. We applaud their success. But what we seem to forget is that they weren't the first to creep on shore...arthropods were.

The best we can figure is that amphibians began the process about 350 million years ago (mya), during the late Devonian period. But long before there were amphibians, long before the first dinosaur left its scaly footprint in shallow seas, arthropods had made that same treacherous leap...from water to land. Fossil records show that approximately 70 million years before the amphibian explosion (i.e., ~ 420 mya), some type of arachnid (a spider or, more likely, a pillbug) had already become terrestrial. The commanding question then, the one we cannot escape, centers on the challenge confronting these pioneers. What did it take to make that giant step that transformed life on our planet?

Keep in mind that most of the earth was under water, and still is for that matter. Keep in mind, too, that the colonizing animal (whatever it might have been) was already adapted for life in the water. Also, we must realize that when the invasion occurred no other animal existed on land; there were no butterflies flirting from flower to flower (indeed, there were no flowers); there were no snakes slithering through the grass, no birds issuing their noisy protest. Land was, indeed, unoccupied by animals. Well, to be honest, there may have been a snail or two sliding through the underbrush. There may have been a few annelids burrowing here or there along the seashores. But truly terrestrial animals? There were none. So I wonder. What compelled those first creatures to leave the comfort of their watery sanctuary to lurch, wriggle, or crawl onto virgin soil?

Figure 12.2 What was the early landscape like, and what did it take for early arthropods (such as centipedes or maybe scorpions) to invade this virgin terrain?

The Attraction……The Challenge

So, what was the attraction? Why did some antediluvian ancestor crawl onto land? In retrospect, we can propose at least three possible reasons: (1) food, (2) an abundance of open niches, and (3) the absence of predators. A combination of those three factors most likely provided an overwhelming pressure to exit an aquatic domain in favor of a terrestrial one. We know, of course, that one more component was involved. Any colonist must have already possessed the appropriate and necessary features to successfully invade land. In other words, the immigrant must have been *preadapted* in some way that enabled it to thrive on land.

However, it is important to remember that in spite of the attractions, terrestrial life is much different than that in water. In many respects, it's much more difficult. To truly become a terrestrial animal, there were several very real impediments to overcome. Among these were:

(1) **Respiration** on land is much different than that in water. Gills, the traditional aquatic respiratory device, doesn't work well on land. First of all, out of water, the gill collapses. As a result, the surface area is reduced below the capacity to support life. Another possible route is to exchange gases directly through the body surface. However, cutaneous respiration, which can work in moist terrestrial environments but not in dry ones, does, by that very factor, restrict the full invasion of arid terrestrial habitats. To overcome these two limitations, the respiratory surface had to be altered: moved internally to remain moist, and kept patent to overcome the tendency to collapse.

(2) Respiratory surfaces are not the only thing subject to dessication. The entire body surface, if not protected, will also dry out if exposed to atmospheric conditions. If an animal dehydrates, it perishes. So, the ability to **prevent dehydration** is another vital step in the total colonization of land.

(3) Because gravity is less of a factor in an aquatic environment, an animal living in water tends to float instead of scraping along the bottom. It doesn't face the degree of friction as does its terrestrial counterparts. An aquatic animal is buoyant. Thus, to colonize land, an animal must overcome friction caused by the pull of **gravity**. Special appendages must be developed that will lift the animal off the ground and, at the same time, propel it forward.

(4) Another advantage an aquatic animal has over terrestrial ones is in the discharge of metabolic wastes and the management of ion and water balance, i.e., **osmoregulation**. In water, an animal can excrete nitrogenous wastes as ammonia which is diffused into the surrounding water. However, on land this is impossible. A land animal must find a way to release metabolic waste, not as ammonia, but in some other form, such as urea or uric acid…what we generally call urine. At the same time, a land animal must balance ion concentrations (especially NaCl) and control water loss. This necessitates a reconstruction of the organ involved (i.e., the kidney).

(5) One of the primary differences between life on land and life in the water is ambient **temperature**. In water, temperature is relatively stable; there's just not much variation. However, on land it's another story; temperature seems to be constantly fluctuating. Shifts occur throughout a given day, season, year, and even within different microhabitats. Life on land is, truly, one of thermal extremes. Thus, a terrestrial animal must be able to compensate for a constantly changing temperature.

(6) All animals require a constant supply of energy. **Acquiring food** posed, at least initially, a problem for land invaders. First of all, since there were no other animals on land during the first wave of colonization a predator would have significant problems finding food. This, of course, changed quickly as animals poured into terrestrial ecosystems. On the other hand, for the first animals, the pioneers, there was an abundance of plant life in the form of ferns, cycads, and mosses. Taking advantage of these new foods posed yet another challenge to land colonization.

(7) To survive, an animal must have appropriate mechanisms to **sense its environment**. The better equipped an animal is to receive and process incoming stimuli, the better its chances of survival. However, waterborne stimuli travel differently than airborne ones. For example, waterborne chemicals are perceived in different ways than those carried by air currents—light travels in distinct pathways on land and in water; and the same is true for touch, sound, and temperature. Thus, to be successful, a terrestrial animal must alter its sensory apparatus accordingly.

(8) Perhaps the singular most important thing an animal does in its lifetime is to **reproduce**. Leaving behind that precious genetic message is an animal's one true legacy. Through reproduction, a populations' inheritable code is made available to the next generation. An important business, indeed. To meet this demand, animals have developed a number of strategies. One that works exceedingly well in water is *broadcast spawning,* the release of large number of gametes into the surrounding environment where fertilization and development occurs. This is the method used by the majority of aquatic animals, although a number of other strategies are also practiced.

There is, however, a real problem with this reproductive strategy on land. To reproduce properly, sperm and eggs must remain moist…they cannot dry out. The same is true for the developing embryo. Thus, external fertilization does not work for land-dwelling animals. To solve the problem of dehydration, land animals have opted for internal fertilization. As a matter of fact, internal fertilization has become the hallmark of terrestrial reproduction.

Internal fertilization requires a pivotal change in reproductive structures and behavior. No longer will spreading one's gametes into the water suffice. Thus, special devices for introduction of sperm into the female's reproductive system must be developed. As a result, a number of intromittent organs (e.g., penises) have evolved. Secondly, like us, each individual animal has a "personal space" that is guarded against encroachment. Since internal fertilization requires that sperm enter the female this cloudlike personal space must be abridged. To do that, animals have developed proper courtship behaviors to break down a partner's natural aggressiveness. Finally, internal fertilization and terrestrial reproduction have demanded a different scheme for development of the embryo. Some animals opt to retain eggs internally (e.g., mammals), while others lay some type of external, shelled egg (insects, reptiles, and birds). Such land-adapted eggs allow embryological development within a controlled, nutrient-rich environment. It is no accident, therefore, that a terrestrial egg has evolved.

The transition from water to land must have been an incredibly challenging and risk-filled venture. But for those animals that achieved it, an unbelievable opportunity awaited. Countless empty habitats were there for the taking. A whole panorama of unoccupied ecological niches spread before them. It wasn't long before those early pioneers gained a toehold and soon they began a massive radiation. They crept, climbed, sprinted, flitted, scurried, darted, tunneled, jumped, and flew in every possible direction until they eventually carpeted the earth.

Animal colonization of land, some 420 million years ago, provided the spark for the diversity explosion that followed, and which continues to the present day. Emancipation from the aquatic environment did, truly, change our planet's landscape.

Barriers to Terrestrialism

1. Must be able to breathe on land
2. Must keep from dehydrating
3. Must overcome pull of gravity
4. Must be able to osmoregulate on land
5. Must be able to deal with temperature fluctuations
6. Must be able to live on terrestrial food sources
7. Must cope with terrestrial sensory stimuli
8. Must be able to reproduce on land

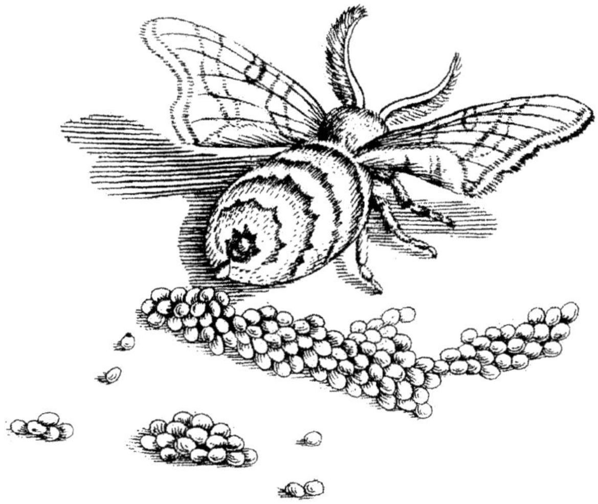

Figure 12.3 To thrive and reproduce on land, animals had to find a way to protect the precious developing embryo without having to return to water. They did this by developing a shelled egg.

Topic 13

Insects and People

Throughout the last 350 million years, insects have exerted an awesome impact on planet earth. Like a ubiquitous whirlwind, they have wielded their massive armies of tiny warriors to influence the biosphere as has no other. All organisms have been affected, and we too, as one of the latest animals to appear, have shared...at times suffering but, in the main, benefiting.

Negative Impact

Pests: There's no denying it, some insects are tormentors. Gnats, cockroaches, houseflies, ants, clothes moths, noisy cicadas, and stinging bees are all bothersome. They take their toll on us, primarily a psychological one.

Parasites: Other insects are even more troublesome since they rely directly on us for food. Mosquitoes, biting flies, bed bugs, lice, fleas, and kissing bugs can be little tyrants as they search our hides for a helping of blood or flesh.

Disease vectors: Still other insects can exact a more devastating price as carriers of disease organisms. Malaria alone, carried by mosquitoes, has killed more people than all human hostilities combined. And it still rampages. Insects also transmit typhus, encephalitis, yellow fever, bubonic plague, anthrax, leprosy, and many other diseases. The common housefly, alone, is vector of over 60 disease organisms

Crop destruction: Each year approximately 10-12% of the world's food supply is either eaten or despoiled by insects. The cost in dollars and human misery is unfathomable.

General damage: Termites tunneling through a house's foundation, carpenter ants boring into wood for a nest, long-horned beetles killing shade trees, clothes moths pilfering a closet of fine garments, carpet beetles fragmenting your new oriental rug, and dermestid beetles turning my bird and mammal museum skins into a shambles are all examples of general damage done by insects.

Benefits

The relationship with insects can be bothersome and, at times, life threatening. They are, indeed, a force to be reckoned with. *But, on balance, they do far, far more good than bad.* The interaction between insects and the rest of us, especially us, is pre-eminently beneficial.

Products: Insects provide us with wax, silk, dye, honey, and even pleasure for collectors who admire their intriguing richness.

Research: The study of insects, termed "entomology," has given us a greater understanding of these organisms. From insect control to basic physiology, particularly, how nervous systems work, entomologists have added significantly to our knowledge of animals. The entire science of genetics is built on the tiny wings, eye color, legs, and antennae of the fruit fly, *Drosophila melanogaster.*

Recycling: Today, in a nearby woodland a deer will die of old age. Or he may succumb to some parasite or to some disease. As he stumbles and falls, the echo is heard throughout ecosystems large and small as millions of his kin, furry, feathery, scaly, or chitinous, do likewise. Have you ever wondered what happens to the legions of carcasses that litter our wildlands and roadways each day? Why aren't we up to our armpits in rotting corpses? Because of insects, that's why. Most dead animals are recycled by insects. Maggots, carrion beetles, wasps, and other insects attack and devour a carcass within days. The value done by this nutrient recycling is inestimable.

Food: Some people eat insects. A chocolate-covered ant gulped at a party or a termite mound torn apart by a nomadic tribe can serve a gustatory function. But think of the nutritional value that insects provide daily to billions of birds, bats, frogs, shrews, skunks, lizards, and snakes. They provide an irreplaceable base to food webs everywhere.

Biocontrol: Insects eat other insects. Sometimes they even eat insects that pester us or eat our crops or destroy our forests. They are nature's own insect control network.

Pollination: But by far the most important contribution of insects is the transferral of plant sex cells. It is estimated that around 65% of all plant species are pollinated by insects. From flowers to towering trees, insects keep the reproductive process going. Without insects we would live in a world of ferns, mosses, and evergreens. Most of our fruits and vegetables are insect pollinated as are our cultivated ornamentals. My flower garden wouldn't exist without visits from bees, beetles, and butterflies.

IT'S A FACT

WITHOUT INSECTS, LIFE ON EARTH WOULD BE A SHALLOW, BARREN LANDSCAPE OF ROTTING, PUTREFYING CARCASSES. TRUE, WE MIGHT BE FREE FROM MALARIA AND YELLOW FEVER, WE PROBABLY WOULDN'T HAVE TO COMBAT PLAGUE OR TYPHUS, WE COULD REST EASILY WITHOUT FEAR OF ATTACKS BY BED BUGS. PICNICS WOULD BE MISSING THE UNINVITED MOSQUITO OR ANTS MARCHING IN CADENCE. WE WOULDN'T NEED TO WORRY SO MUCH ABOUT CROP DEVASTATION; BUT THEN, THERE WOULD, OF COURSE, BE FEW CROPS. WITHOUT INSECTS OUR FOOD SUPPLY WOULD DWINDLE TO VIRTUALLY NOTHING. OUR WORLD WOULD TRULY BE IMPOVERISHED BEYOND RECOGNITION WITHOUT THESE TINY SWARMING... SUCCESSFUL...ARTHROPODS.

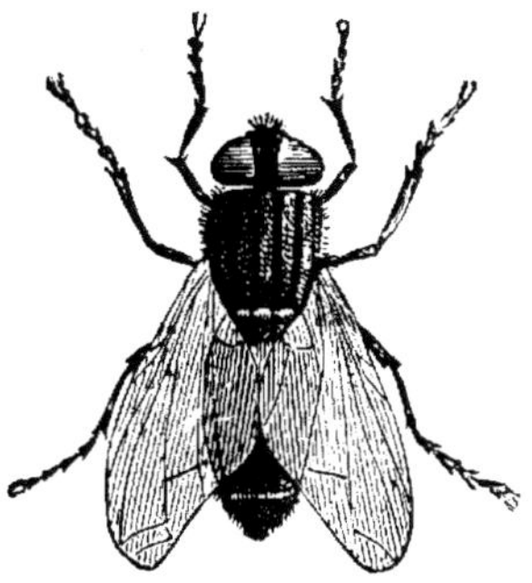

Musca domestica
The Common Housefly

• Houseflies are hardy, ubiquitous pests found wherever humans live. Although they cannot survive temperatures below freezing, they can, because of their close association with people, overwinter in cracks and crevices in our homes. In fact, because of central heating, they can, at times, breed the year around right beside us.

• Each compound eye contains approximately 4,000 lenses; thus, flies have a wide-angle, omnidirectional visual field. That's why it's so difficult to swat one.

• Unlike most insects, flies only have one pair of wings; the rear ones are reduced to club-shaped structures, the halteres, which help to maintain balance while flying. The wings beat at a speed of almost 20,000 beats per minute (that's about 330 beats per second). Compared to the hummingbird's pace of 80 beats per second, the flutter of fly wings is a virtual blur!

• Their food consists primarily of sugary materials, although females feed on protein when forming eggs. Prior to feeding, houseflies regurgitate (i.e., vomit) on their food to liquefy it, and then sponge up the sugary broth. On their tiny feet are special receptors that they use to taste potential food.

• Houseflies have the habit of defecating every four to five minutes, usually on their food.

• Copulation, in houseflies, takes approximately one hour. After mating, the male injects a fluid into the female that makes her unwilling to mate with any other male; thus, he assures that his sperm will be the ones to fertilize the eggs. Following copulation, the female deposits eggs in other animals' excrement, in rotting carcasses, or on garbage.

• Development, from egg to pupa to adult, occurs in about 18 days. Adults normally have a life-span of approximately 30 days, but may live as long as 5 months.

• Houseflies have an enormously high rate of production: a single batch of 120 eggs from one gravid female would, in five months, yield about 191,000,000,000,000,000,000 flies if they all survived. Some experts claim that this equals enough flies to cover the earth's landmass to a depth of 18 feet! In point of fact, more flies hatch each day than there are humans on earth.

• The housefly is a vector of approximately 60 disease-causing organisms such as typhoid fever, leprosy, gangrene, bubonic plague, dysentery, tuberculosis, anthrax, and cholera.

• They may even spread ulcer-causing bacteria. Most of these disease organisms become attached to the feet of the fly following a visit to feces, carcasses, or other waste material. The disease organisms may then become transferred to human food as the fly is feeding. Houseflies can also harbor ~ 6 million bacteria on the outside of their body, and another 40 million internally.

• Just a few decades ago, houseflies were susceptible to pesticides such as DDT; however, they are now highly resistant to this and many other chemicals used in our battle to control household pests.

adult fly

larva (maggot) in a puparium

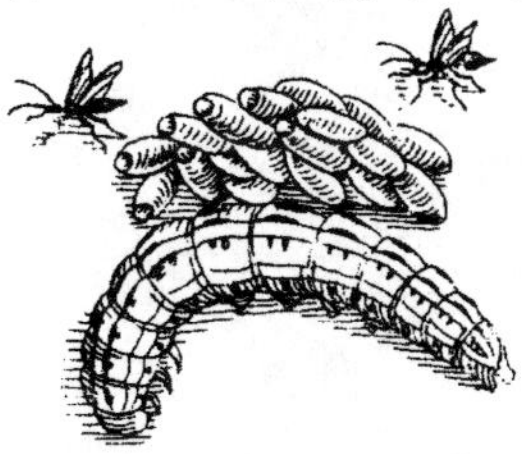

eggs on a dead caterpillar carcass

Topic 14

Symbiosis

Big fleas have little fleas upon their backs that bite 'em,
and little fleas have lesser fleas,
and so on ad infinitum.
- Anonymous

Introduction

In nature, animals never live in total isolation; there are no hermits in the world of animals. A single, unattached individual just doesn't exist. Instead, animals show a vast facility to interrelate with each other. In many ways, and on many levels, they associate together. Often the interactions are so intricate to be beyond imagination.

Sometimes the relationship is required, or **obligatory**, in that one or another participant cannot survive without it. It is necessary for survival. An example would be a tapeworm and its host. The tapeworm is incapable of surviving without that special relationship. At other times the connection is not essential but is exploited if, and when, the opportunity presents itself. This type of relationship is known as a **facultative interaction**. Some leeches, for example, will assume a parasitic lifestyle and feed on blood if given the chance. But they can live very well on their own as carnivores or scavengers.

Another way of viewing animal interactions is to consider the kinship of the participants. If an interaction occurs between members of the same species, it is then an **intraspecific relationship** (e.g., male-female contacts; parent-offspring relationships; play behavior; group foraging efforts, social grooming, and so on). On the other hand, the relationship may be between members of different species. This latter interaction, an **interspecific relationship**, is what concerns us here, for, in biology, the term **symbiosis** is used to identify *any long-lasting, intimate association between members of two or more species*. In other words, symbiosis is a close, enduring interspecific relationship.

Such interspecific relationships are widespread. In fact, virtually all animals serve as a natural habitat (**host**) for hundreds, if not thousands, of other organisms (**symbionts**). The interactions can also be complex and, at times, bizarre. Nonetheless, although wide-ranging, symbiosis, seems to follow only four patterns.

(1) **Mutualism**: There is a special relationship that exists between termites and their intestinal protozoa. These gut parasites break down cellulose as a food source and in doing so release nutrients for their termite host. Without their microscopic hitchhikers termites could not survive on their wooden diet—and, of course, neither could the protozoa. In this case, and countless others, a *mutualistic* relationship has been formed and both the symbiont and the host benefit. These associations provide advantages

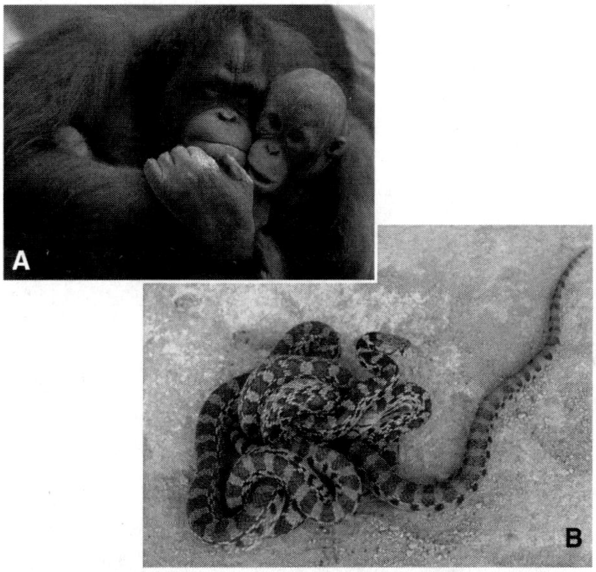

Figure 14.1. Intraspecific relationships occur between members of the same species such as the strong bond between mother and infant (**A**) or the interaction that occurs during mating as in the snakes above (**B**). These are *not* examples of symbiosis, which, instead, refers to interspecific interactions between members of *different* species...the subject of this topic.

such as housing, protection, food, or some other mutually beneficial resource. Mutualism is a common type of symbiosis and is often the result of complex adaptive interdependence. That is, both parties adapt to each other throughout the association and, in fact, **co-evolve** in response to the relationship. As one party becomes better adapted to the interaction so, likewise, does the other. (see Topic 20)

Examples of mutualism are many; here are but a few: photosynthetic algae (zooxanthellae) provide an energy source for corals while benefiting from a protective habitat; sea anemones hitch a ride on the back of certain crustaceans, thus allowing the otherwise sedentary anemone a chance to relocate, while the anemone's stinging tentacles provide the crustacean with some protection from its enemies; gleaning fish and shrimp remove ectoparasites from a variety of marine fish; oxpeckers, likewise, remove external parasites from many large mammals; and a species of ant serves as bodyguard to a caterpillar who, in turn, awards drops of sweet nectar to the waiting attendants.

(2) **Commensalism**: This is a somewhat less common association in which the symbiont is benefited while the host remains unaffected. Remoras are small fish that attach to sharks and other larger fish to forage on leftover scraps of food...they profit while the shark is not affected. Certain fish and crab species receive protection as they hide among the sharp, toxic spines of sea urchins; the urchin is indifferent. Many fish cruise along with large predators receiving food scraps along the way.

There is one special example of apparent commensalism that is somewhat confusing. An uncertain relationship exists between several species of clown fish and venomous sea anemones. The fish live among the anemone's stinging tentacles, but it is unknown if the anemone gains from the relationship. Perhaps the clown fish remove debris from the tentacles; if so, then the relationship would be more likely one of mutualism.

(3) **Phoresis**: In this simplest of symbiotic encounters the symbiont uses the host only for transport and nothing else. A wading bird may inadvertently pick up fish, insect eggs, or larvae and transport these to another locale. Barnacles affix to the surface of whales or other large

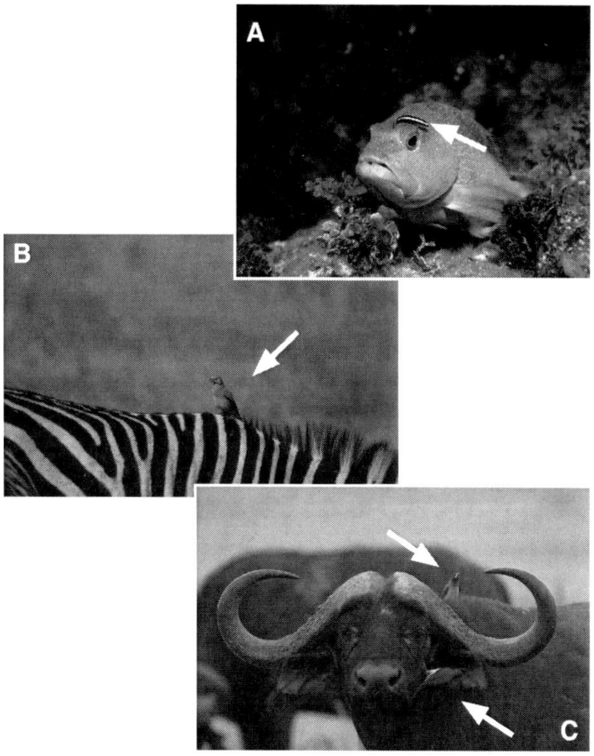

Figure 14.2. Mutualism is a common interaction between members of different species in which both parties derive a benefit. Some symbionts are adapted to pick ectoparasites from their hosts as in the cleaner wrasse (**A**) and oxpeckers (**B** & **C**).

Figure 14.3. Remoras (**A**) have a modified fin to attach to large fish. They most likely receive scraps of food as the larger fish feeds, as do the small fish in plate (**B**) shadowing the hammerhead shark. Some fish can use the spines of sea urchins to protect them from attack (**C**). Several species of clown fish (**D**) live within the stinging tentacles of sea anemones. It is suspected that a mucous covering keeps the fish from being stung. It is unlikely that the anemone derives any benefit.

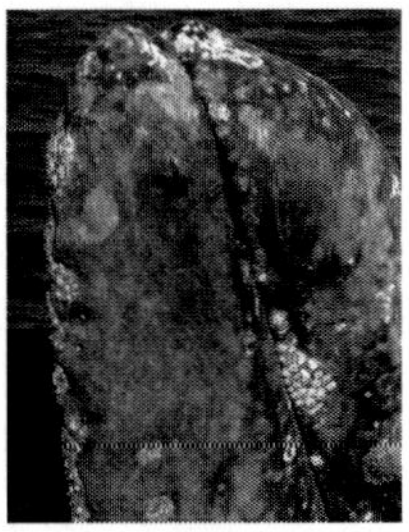

Figure 14.4. Barnacles demonstrate the phenomenon of phoresis by hitching a ride on the surface of whales.

marine animal...they get a ride and the host is untouched. In simple terms, phoresis involves one organism using another to "hitch a ride"; thus, the symbiont is dispersed into new surroundings. Although the phoront derives some benefit from the relationship, its survival does not depend upon it. And, of course, the host is not affected. In this regard, phoresis is similar to commensalism.

(4) **Parasitism**: There's little doubt that the most common symbiotic relationship among animals is parasitism. Here the symbiont (parasite) *benefits at the expense of the host*, usually by gaining some metabolic advantage. Sometimes, in fact, this relationship is referred to as "destructive symbiosis." In one way or another the parasite exploits the host to the latter's detriment.

However, while the host is adversely affected it is usually not critically harmed. It is to the parasite's advantage not to seriously injure or kill its host, for to do so would result in the parasite's demise as well. In parasitism, it seems that the longer the relationship, the more benign it becomes. **Co-evolution** apparently works its genetic magic to allow the two to live in relative harmony. But there are many exceptions, and trying to understand exceptions can be challenging.

Generally, in pathogenic cases wherein the parasite causes severe disability or death, the relationship is considered to be of recent origin. An example can be cited between large African mammals and a species of *Trypanosoma*, a blood protozoan. In this instance, wild hosts are not seriously affected by the parasite but when domestic cattle were introduced into the region, they became infected and, as a result, large numbers of the cattle died. Those that survived will pass on the survival trait, and, over time, a state of relative accord should develop.

Parasites wear many costumes. Those that occur on the host's exterior (*ectoparasites*) are modified in different ways than their internal cohorts (*endoparasites*). Whether occurring internally or externally, the parasite is specifically adapted to gain safe refuge, food, reproductive opportunities, and other advantages. In fact, the parasite is often so intricately adapted that it can reside only in certain specific locations within or on the host, and nowhere else. Examples of **host** and **tissue specificity** of this sort are many: *Fasciola hepaticais* a fluke confined to the liver of a sheep; *Clonorchis sinensis*, another fluke, is only found in human livers; tapeworms, likewise, are limited to the intestine of specific hosts; heartworms to the heart cavities of dogs; lungworms reside only within the lungs of frogs; and so on. There are even larval mollusks, baby clams, that parasitize only the gills of certain fish.

Whatever you might say about parasitism, one thing is certain...the parasitic lifestyle is extremely popular. Of the 34 animal phyla, 14 contain parasitic members, and 5 are entirely parasitic. From protozoa to insects, from fish to bats, the parasitic stage has seductively lured its players, and the results can be hideous: a catfish that lodges inside of the human urethra; roundworms that cause the ghastly swellings of elephantiasis; tapeworms that cause blindness; mites living in the ear of a moth; a protozoan that slowly dulls the mind in African sleeping sickness. There are even worms that romp in the sinuses of narwhals, driving their host nearly insane with their endless cavorting.

Figure 14.5. Virtually every animal has parasites. Here one baboon removes ectoparasites from another in an act of social grooming. In addition to the external parasites, there's little doubt that both animals also harbor a number of internal parasites.

Topic 15

The Beauty of Parasitism

Every known vertebrate animal (and most invertebrates, as well) supports an entire congregation of parasites. Flukes, ticks, mites, protozoa, fleas, lice, tapeworms, bedbugs, leeches, biting flies, roundworms, lampreys, and even vampire bats are members of this exclusive fraternity—each with its own special needs; each with its own special contraptions, modifications, and adaptations.

One of the things that makes the study of parasites so fascinating to the zoologist, and anyone else who looks closely at the parasitic relationship, is that the "creativity" of parasites is simply astonishing. Yes, they can be repulsive vagabonds from our viewpoint as potential victims, but they are also, without question, awesome creatures. Their specialized adaptations are simply anatomical and physiological wonders. The degree to which parasites have exploited evolution's promise is nothing short of spectacular. They are grotesquely beautiful examples of evolution's power to invade the most unimaginable niches.

Parasitism: What's the Attraction?

The question then becomes, why so many? Or stated another way: What is the lure that draws such an ungainly gang to this strange existence? Well, the answer is not too difficult to find. Even a casual examination of parasitism reveals three possible forces that might cause a free-living animal to become fettered to this lifestyle. But three powerful, magnetic forces they are: *food, safety, and sex.*

There would be little argument in the scientific community that the primary motive of parasitism is a readily available food supply. Whether it's a blood meal for a mosquito, flea, tsetse fly, leech, or vampire bat, the attraction is powerful. Likewise, the cell debris, tissue fluid, and mucus meals of flukes and roundworms is equally tempting as is the predigested paste of the tapeworm terrain. Lice and mites apparently get excited over scalp flakes, while tiny catfish go ballistic over urine sediments. Food is, indeed, a compelling excuse for the parasitic way.

And to this just add the fact that the banquet can be served in a safe, secure hideaway, free from predators, free from extremes of climate, free from excessive competition. To have a place where food is plentiful, easy to obtain, and where life is secure must have been a truly powerful attraction.

Then, as if plenty of food and a safe shelter weren't enough, along comes one more temptation: ready and willing reproductive partners. In the everyday life of many parasites, there is an ample supply of potential mates, indeed, you may be surrounded by them. Parasites don't need to spend time and energy seeking out partners; their kitchen and bedroom are one in the same. Surely, the availability of reproductive partners further spurred the rush to enter the parasitic lifestyle.

These three, ready food, a safe harbor, and accessible partners must have combined to present an overwhelmingly tempting new life. So attractive are these three forces that they served as an invitation to thousands of free-living animals to forsake an independent existence in favor of willing bondage. So, just maybe we're asking the wrong question. Instead of wondering why there are so many, many parasites, perhaps we should marvel at why there aren't more!

Is It Really That Easy?

The answer is no. Attractive, maybe; easy, no. Becoming a successful parasite isn't without its share of difficulties. In fact, there are a number of very real barriers that parasites must overcome in order to take advantage of the Utopia that awaits them.

(1) **Finding the host.** Just locating the proper host can be a major hurdle for a parasite. But they can be extremely inventive in circumventing this obstacle. For instance, parasites have discovered that by linking to the host's food they stand a better chance of ending in or on that host.

So we find tapeworms ensconced in beef, pork, rabbit. We find flukes in fish, in a snail tentacle, or on edible vegetation, and we see roundworms buried in the flesh of a pig. Furthermore, parasites have recruited a number of insects to join their search team: mosquitoes, biting flies, fleas, lice, ticks, and many others all serve as active transporters (vectors) of parasites.

(2) **Attachment in the right place.** However, finding the proper host is only half the battle. Parasites must then pinpoint their favorite, specific tissue and once there they must stay put. Parasites have, therefore, invested much of their evolutionary effort toward attachment devices. They are adorned with hooks, suckers, spines, hairs, glue, claws, teeth, and even lips to vigorously cling to their secret places.

(3) **Overcoming host defenses.** So far the attack seems to be one-sided, all in the favor of the parasite. However, that is far from the case. Hosts are not passive recipients in the game of parasitism. Indeed, they have an entire battery, a virtual armory of defenses at their fingertips. Acidic intestinal fluids, an anaerobic milieu, high temperatures, and explosive muscular contractions are all forces opposed to parasitic survival. To be a successful parasite, these defenses, too, must be conquered.

The inside of an intestine is pretty inhospitable. It is a dark, hot, acidic, anaerobic, and pulsing environment. But as stark as the intestine might be, it's not the front line of defense against parasitic attack. No, that honor goes to the immune system. Imagine, an entire body system designed just to remove foreign material, just to repel tiny invaders. If the immune system recognizes a parasite as an alien intruder, the parasite is, in most cases, doomed. Thus, the parasite must disguise itself as friend rather than foe, or else face inevitable eviction. There are many parasites that are quite adept at doing just that. Through a process called **molecular mimicry** some parasites can imitate secretions or other chemical features of host tissue. By doing this they con the immune cells into accepting them as a natural part of the host rather than as invaders.

(4) **Surviving outside of the host.** For a parasitic animal to be really successful, it must spread from one host to another. That means that it must survive, for a period of time, outside of the final host. The adult parasite, or one of its larval stages, must find a way to leave one host and then survive long enough to re-enter another one. Parasites have, for the most part, chosen to do that in immature states: eggs, larva, or juveniles.

The methods used are numerous and often farfetched, yet they fall under only three general strategies: (1) massive egg production; (2) endurance; and (3) the use of intermediate hosts. Parasites are legendary for their prolific egg production. A single female human roundworm can issue 200,000 eggs per day, during her 1-year lifetime. Not to be outdone, tapeworms can release one or more gravid proglottids daily, each containing upwards of 100,000 eggs. It seems that this strategy recognizes the nearly impossible survival odds and responds by flooding the market with tiny, host-seeking vagrants.

The second survival strategy involves an almost unbelievable capacity for tolerance. Both eggs and larvae can withstand the harshest of conditions. They can become drought or temperature resistant, can survive without food, or even resist abrupt changes in pH, and often they will become dormant. In a state of dormancy, the parasite reduces its metabolic activities, shrivels up, and just waits. Their patience is astounding, since the wait can be a long one. Parasites have been known to re-emerge from a cryptobiotic state decades after entering, none the worse for wear. There is even considerable anecdotal evidence of much, much longer dormant periods.

But it's the final strategy that truly stretches our credulity. How are we to believe that a parasite located in our innermost recesses, say our blood vessels, can enlist the aid of a tsetse fly to transport it to another unknowing victim? Consider these lengths to which some parasites reach to use insect bloodsuckers in their quest to locate another host. There's a nematode, *Wuchereria bancrofti,* that lives within human lymph nodes causing the hideous condition of elephantiasis. It produces a larva that lives, during the daytime, within the deep blood vessels of the body, but which travels each evening toward the skin where mosquitoes are feeding. The parasite actually synchronizes its behavior to that of a mosquito it has never met and cannot see.

It's beyond belief that a parasite can even change the behavior of another animal to improve its own chances of survival. But there are, in fact, many examples of just that. There's a species of snail that becomes so confused by the presence of parasites that it exposes itself to possible bird predation. And there's a fish that becomes so disoriented by the presence of internal parasites that it fails to flee oncoming predators—predators that will become another host for that parasite.

Then there's the little fluke, *Dicrocoelium dendriticum,* whose larvae invade ants. The infestation causes the ants to forego normal secretive behavior to, instead, plunge themselves upward to clutch blades of grass with their powerful jaws. There they hang like miniature Christmas tree ornaments until along comes a herbivore, leisurely grazing and unwittingly ingesting a little parasitic surprise along with its grassy feast. Truly, parasites have left no leaf unturned in their greedy pursuit of one host after another.

Problems Faced by Parasites

1. Finding the host
2. Finding the right place in the host
3. Overcoming defenses of the host
4. Surviving outside of the host

Topic 16

Some Human Infestations

It is unfortunately true that we are not exempt from this parasitic battlefield. As a matter of fact, humans are equal opportunity employers when it comes to engaging parasites. We have our share of critters crawling over our skin, seeking special treats at the base of our hair follicles, looking for private sanctuaries from which to withdraw a sip of blood, creeping over our surface devouring fleshy flakes, gushing within our blood vessels, hanging about our intestines, lurking within our livers, or seeking refuge in our muscles, skin, eyes, lungs, urinary tract, and so on. Yes, we have our share, and while most are relatively harmless, some do exact a punishing toll.

Leeches and Rattlesnakes

How do they find me?

For each of the past 22 summers, I have taken an odyssey to the Boundary Waters Canoe Area on the border of Minnesota and Canada. There, in the great outdoors, I seek nature's antidote to the urban murrain. I escape into a world of eagles, campfires, purple sunsets, and loons. However, on a number of trips I have had the misfortune of meeting another presence, a miniature vampire of sorts. For on many occasions, I have served as leech-lunch. I have found them securely fastened to my arms, legs, and other abstruse locations. Without exception I am always surprised at how quickly they find me. It seems that no sooner do I step into a marsh, a pitch-black lake, or a quaking beaver pond than I see a shadowy clump hunkered down on my calf or arm. They seem to arrive out of nowhere with me as their hapless target. Just how can they be so expeditious?

Well, the truth is, hungry leeches have at their disposal various means to locate a warm-blooded host. First of all, they have, scattered over their body, special tactile sensors able to discern slight vibrations in the water. So, by the time I step into the cold darkness, they are already on the alert. Then, by using thermoreceptors, they are able to sense my body heat. I guess even on the coldest, rainiest day, I am still cuddly enough to merit their celebrated kiss. Finally, they smell me; and after a week in the back country, I'm not too surprised. They have epidermal chemosensors that detect substances oozing from a potential host. As a matter of fact, you can perform a simple test to demonstrate their sensitivity to faint human secretions. Just touch the inside of a beaker containing hungry leeches. They will immediately detect the oils on your fingerprint, become agitated, and quickly move toward that spot seeking a bloody meal. So, how do they find me? That's easy; to a leech, I am a huge, warm, vibrating, smelly target.

Why do I tolerate them?

Frankly, I endure them because, for the most part, I don't even know they're there. From the leeches' viewpoint, once they've located a meal, they must be able to attach and feed relatively unmolested. But how can they extract blood, of all things, without being detected? To do so, they have a couple of tricks up their slimy sleeves. First of all, leeches have three razor-sharp teeth that can quickly slice through flesh. This they can do unannounced because they have, in their saliva, an anesthetic that numbs the immediate area.

Thus, I can be attacked without my knowledge, often by a relatively large number of leeches. In addition to the anesthetic, they also inject an anticoagulant, which delays blood clotting, sometimes for up to 48 hours. And if that's not enough, they have one more secretion, a vasodilator that expands my small blood vessels. These three salivary secretions allow the leech to gorge itself with

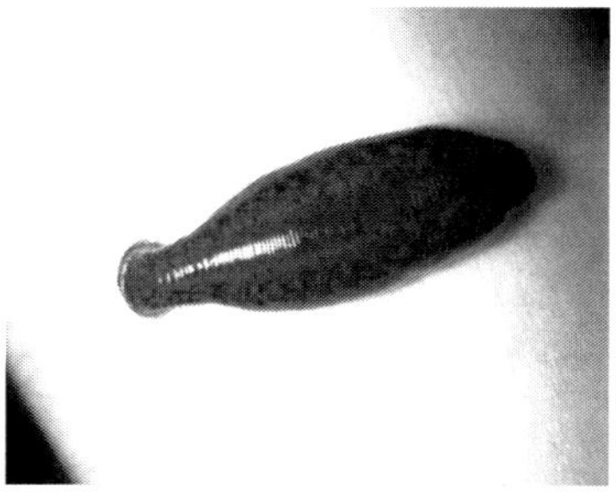

comparative ease, and without much interference. After a complete blood meal, lasting from 10 minutes to an hour, the sated leech drops off, not to feed again for about 12 months. That's a relief... but come to think of it, that's just about the time I'll return for my annual getaway.

Medicine's Bloody Little Secret

Leeches and medicine have coexisted for a long, long time. The earliest records out of ancient Egypt refer to the medical use of leeches. I'm certain that if records existed from our cave-dwelling ancestors they would tell stories of the power of leeches.

The reason is linked to the early interpretation of disease. For most of our history as a species, it was thought that diseases were due to an imbalance in the body's "humors." In other words, the medical community believed that human afflictions were due primarily to the presence of "bad blood." The remedy for such a simple diagnosis was equally uncomplicated. Just remove the offending fluid, bloodletting it was called, or more precisely, leeching. Barber-surgeons (that's right, barbers; the traditional red-and-white barber pole reflects their one-time status as surgeons!) would just attach several leeches to the body and drain away the foul blood. Virtually any human ailment was thought curable through bloodletting. Fevers, aches, obesity, tumors, boils, laryngitis, acne, pneumonia, nausea, headaches, and even mental illnesses were among the afflictions for which leeches were the favored treatment.

Bloodletting was so prevalent that a thriving industry of leech collectors developed. Collectors simply walked slowly through infested waters and then picked the feeding leeches from their legs or torso. During the 18th and 19th centuries, in Europe alone, millions upon millions of leeches were used annually. Collection became so intense that the number of leeches finally dwindled and, in fact, in France and England, they virtually disappeared. Even today the leech is an endangered species in England. The practice of leeching peaked in the 1830s and then fell into disrepute as the medical world began to uncover the real causes of disease. By the end of the 19th century, leeches were all but eliminated from the physicians little black bag.

However, one question remains...did it work? Did bloodletting really help the patient? What do you think? Would draining a soda-can full of your blood cure your headache, resorb that pesky pimple, or take off a few pounds? The fact is, bloodletting offered little relief to the ails of our forefathers. On the contrary, leeching, in many instances, gave the patient a bitter dose of anemia, or even worse.

A note of interest here is that George Washington, after contracting a severe sore throat, was bled three times on December 14, 1799, although with instruments, not leeches. He died that very day. While any connection between bloodletting and his death is speculative, it's quite obvious that bleeding sure didn't help. Another interesting side note is that it wasn't always the patient who was victimized. It is well known, for example, that leech collectors, too, often became pasty-faced anemics.

The ebb and flow of medicine is, at times, strange to behold. What is common practice one day, is phased out, only to return later as a promising new elixir. Leeches are a good example. They were, for most of human existence, a popular treatment for our maladies. Then they lost favor, their use became ridiculed and subverted.

But they were down, not out. As a matter of fact, they have recently resurfaced; they have again raised their thirsty little lips to re-enter the world of medicine.

The modern physician recognizes that *Hirudo medicinalis*, the medicinal leech, can, indeed, be an ally if used properly. They are no longer applied indiscriminately to relieve a patient of a cup or so of blood. Today, they are used quite specifically. For example, in those cases where blood becomes pooled just beneath the skin, such as around the site of a suture, or even a blackeye, leeches can be quite effective. By placing a few hungry ones near a subcutaneous wound, the blood can be safely and completely removed without the need for further surgical intervention. Also, as the leech feeds it injects *hirudin*, a powerful anticoagulant that is helpful in restoring blood flow following reconstructive surgery.

Because of hirudin, leeches may help in reestablishing blood flow following attachment of severed fingers, hands, ears, or even dissevered penises.

They are also used to treat varicose veins, cauliflower ears, and glaucoma as well as aiding in the bonding of tissue grafts. Lately, leeches are being used to prevent, or even dissolve, blood clots. In fact, researchers have recently isolated compounds in leech saliva that show promise in the treatment of cardiovascular disease.

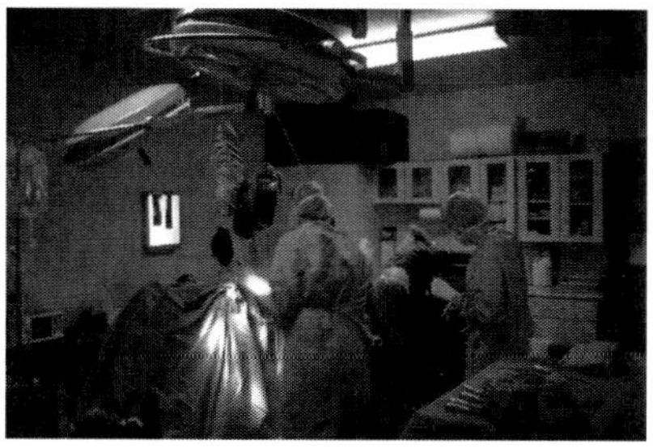

Snakebites, Anemia and AIDS

Leeches have also been used to treat snakebites. The venom injected by poisonous pit vipers, such as a rattlesnake, is a complicated cocktail of toxins and proteins. A snakebite may evoke a number of symptoms, usually involving tissue deterioration and subcutaneous hemorrhage at the injection site. In some cases, leeches may be effectively used to remove blood collected beneath the skin following snakebites.

There are, however, some risks involved in using medicinal leeches. Anemia is experienced by approximately 50% of patients following prolonged leeching. And the physician must take care that leeches do not roam. For instance, if they are applied to the facial area and are not sufficiently confined, leeches may migrate into the nasal passages or bronchial airways causing serious respiratory blockage or dysfunction.

Then there's the question of HIV and AIDS. While leeches have never been implicated in the transfer of HIV, the possibility does exist. Thus, leeches that have been used in a medical procedure are discarded as biohazard waste; they are never reused.

In spite of these few side effects, the leech, like Phoenix, has risen from obscurity, renewed and reinvigorated, to resume its place among medicine's ever-expanding arsenal of MRIs, CAT scans, lasers, electron beam scanners, laparoscopes, and heart transplants.

Mighty Mites

...the average living room is a hive of activity and the bedroom is probably busier still,
where even the most virtuous of us share our beds with thousands of other creatures.

- Brookesmith

Facial Warriors

My guess is that this morning you bathed. Or, perhaps it was yesterday? Maybe you used one of those liquid, antibacterial soaps; after all, the primary aim of bathing is to rid ourselves of external microbes and the smell they spawn, body odor we call it. The problem is that you, like everyone else, probably weren't very successful in combating these ubiquitous foes. But rest easy, you have a comrade, an ally, in this fight. Unknown to you, day in and day out, your skin is a battleground where tiny warriors contend for supremacy. Bacteria are feeding on your surface debris and mites, in turn, are feeding on the bacteria. That's right, mites, tiny arachnids, minuscule parasites are in residence on your body's surface. You have them, I do, we all do. They are skin mites with the impressive names of *Demodex folliculorum* and *Demodex brevis*.

The first type (*folliculorum*) resides in pits at the base of hair follicles, and, for some reason, they prefer the hair of your eyebrows. The other species, *brevis*, lives in the adjacent sebaceous glands. Both like to gobble up skin flakes, body oil (sebum), makeup, or anything else lingering on your wrinkled brow. But what they really, truly relish are bacteria. These mighty combatants escape the friendly confines of hair follicles or sebaceous glands, to live for a few days on your face, feeding on the abundant bacteria, and then crawl back into their crypts to lay eggs. The newly hatched mites eventually make their way to the surface, as their parents did before them, and the whole cycle repeats.

However, unlike some parasites, skin mites are harmless and, in fact, may actually contribute to the health of your skin! But where did they come from? How did you become contaminated with these miniature hedge hogs? Well, it seems that you can pick them up in a variety of ways, usually through contact with an infected person. They are transferred by casual touching, by wearing someone else's hat, using a friend's face cloth, or perhaps through kissing. There's even evidence that good ole mom may have delivered them to you during breast feeding! And just in case you would like to view these miniature gluttons crawling about your face, here's how. Pluck a few hairs from your eyebrows, place them on a microscope slide, and take a peek under low power.

You will discover that they are quite cute. Like little stubby cigars, these tiny mites, with tiny legs, patrol the cosmic moonscape of your forehead searching for a feast of bacteria and dead skin. (Fig 16.1)

Pillow Fights and Dust Storms

One day not too long ago, while reading near a window, I became captivated by the flow of dust particles streaming in the sunlight. I knew that people of science were supposed to be inquisitive so, in keeping with that pioneer spirit, I began to contemplate dust. Just what is it? Where does it come from? What's it doing in my house? Why is it streaming in my sunshine? Well, after some searching and rummaging in one dust-covered book after another, here's what I found out. First of all, about 80% of household dust is dead human skin. That's right, skin, our own skin. After some rather unsettling reflection on that little fact, I finally concluded that it's really not too surprising since skin is our largest organ and it's replaced every month or so. New cells, by the millions, constantly, incessantly, push dead ones up to the surface where they fall away in dusty little clouds. Imagine, millions of dead, scaly, skin flakes shed by each of us every day. They take up an airborne existence to travel around our homes, offices, schools, and restaurants like tiny snowflakes drifting aimlessly here and there. I discovered that they float on currents of air outward from the center of a room toward its edges where they fall in a veritable blizzard. That's why dust concentrates on picture frames, fireplace mantles, TV sets, and unopened biology books.

If roving skin makes up around 80%, what's in the other 20%? What's lurking there? Experts in dust content tell us that there is a patchwork of items in that remaining one- fifth. It seems, for instance, that in addition to skin flakes, dust also contains a share of pollutants, pollen, human hair, carpet fibers, insect and spider excrement and their exoskeleton fragments, and, oh yes, the mites. Carpet mites, otherwise known as *Dermatophagoides pteronyssimus,* and these, even more than their *Demodex* cousins, relish dead skin. They drift from floor, to ceiling, to window sill, to grandfather clock, in a constant tide of dust, feeding all the while on those gliding flakes of skin. Come to think of it, I'm rather glad they're here among us. Who wants to wake up each morning covered by a film of dusty scales?

Now that brings us to the bedroom. You would think that there's at least one pest-free place in our lives, one sanctuary where we can escape Mother Nature and her mischievous moods. Although there may very well be such a place somewhere, it's not your bedroom, not your bed, and certainly not your trusty, cherished pillow. In fact, a bed is a perfect incubator for mites; it's warm, it's humid, and it has a nocturnal, endless source of food. It is estimated that a single bed is a breeding ground for millions of dust mites. There we emit our body heat, breathe and sweat moisture into the mattress, toss and turn constantly scraping off dead flakes like falling snow, all the while providing nourishment for our tiny bed companions.

Oh, there's one more thing. About that pillow of yours, and who doesn't like to have a nice fluffy pillow on which to rest one's weary, fact-filled head? Pillows, one of life's true comforts. But, you guessed it! Mites adore pillows; it's their Garden of Eden. Hundred of thousands tumble out when you "plump" one, and washing doesn't seem to offer much control. There, on our warm pillows, mites feed, defecate, lay their eggs, raise their young, and die. And while they don't seem to harm us any (unless you suffer from asthma), somehow knowing they are there makes me a bit uncomfortable. Somehow, my psychological wellbeing doesn't relish the idea of sleeping on a mat of mites, rolling about in their dead carcasses, wallowing in their copious excrement. As a matter of fact, it bothers me a lot. And I suppose the next time I drift off to sleep in that hotel room, I'll be thinking of the many squatters who have preceded me offering their flaky feast to the legions of mites nestled somewhere below, keeping them fat, healthy, and happy.

The Shameless Tormentor *Pediculus humanus*

Having another lousy day? Are you surrounded by general lousiness? Do you louse up from time to time? Was your professor just too nitpicky on your last zoology exam? Truly, lice are ingrained in our human experience. They have been inseparable partners with us for a long, long time. Probably as soon as we gathered into bands of Neolithic foragers, we became home for these six-legged tormentors. And an encounter with *Pediculus* is just one more reason to feel lousy. In fact, fighting off a louse infestation can be an unpleasant, embarrassing, and a rather lopsided affair; all tactics seem to favor the louse.

In his marvelous little book *Rats, Lice and History* Hans Zinsser places lice-borne typhus among the true scourges of mankind, right up there with bubonic plague, malaria, leprosy, small pox, and war-time casualties. He recounts the many creative ways we have tried, for thousands of years, to rid ourselves of this pest. So you would

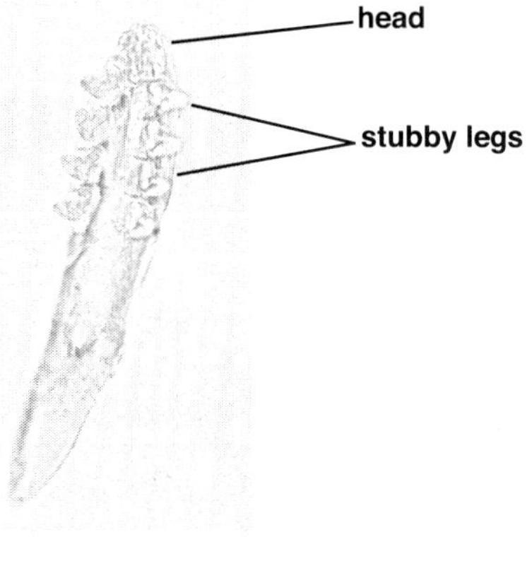

Figure 16.1. The cigar-shaped skin mite, *Demodex*, lives in association with hair follicles where it feeds on skin flakes and bacteria. The photograph to the left shows a skin mite attached to a hair shaft.

think that such a nuisance would by now be pretty much eradicated. Right? Of course not. Indeed, it seems that an honest-to-goodness present day outbreak is among us. Estimates place the number of Americans infested at around 6–12 million. Most of these are children under the age of ten.

Public health records indicate that the infestation is growing with three reasons generally cited:

(1) a sharp increase in the number of preteen children,

(2) overcrowded schools, and

(3) the emergence of a "super louse," resistant to modern-age chemicals.

That sounds just great...a super louse. The plain old louse of our ancestors wasn't enough; now we have our very own creation, and it's bionic!

One of the factors that make lice so hated is the ease with which we contract them. Just casual contact with an infected person, his clothes, or personal belongings, can transfer eggs or adult lice to me or to you. And while infestations are often correlated with crowded, filthy, impoverished conditions, lice are really not all that particular. They infest the wealthy along with the poor, the clean and the unkempt, the socialite and the homeless, the urbanite and the suburbanite.

To make matters worse, as easy as it is to acquire lice, it's even easier to keep them. They have a pure affection for our hairy domes and once ensconced among blondes and brunettes, redheads and silverheads, they prefer to stay. They are efficient, enterprising, and, yes, loathsome creatures that can do four things very well: suck blood, stay put, glue eggs (called nits) to hair shafts, and cause general humiliation. No parasite elicits the sense of repulsion like that of a head full of lice. If you can get anyone to admit to an infestation, you will be told of the shame, the stigma, the embarrassment, and the sheer humiliation. So, it's not bad enough that they plunder our scalps, use our blood as an unending reservoir of food, and incubate baby nits among our curls, they add a healthy dose of insult as well.

But humans are a resourceful and persistent lot. We fight back. In the U.S. alone, there is a multi-million dollar industry built up around delousing. Stainless steel combs designed just to remove the nits are available at any drugstore, or any decent web site. Great Britain is so infested that the country has a semiannual "Nitpicking" day, everyone's included.

The truth is, lice are difficult...no, very, very difficult to control. There are several home remedies that are virtually worthless (Vaseline, kerosene, alcohol, baking soda, gasoline). Both over-the-counter and prescription medications are available to help control lice infestations. However, the only certain method is the tedious removal of the rice-sized adults or the even smaller nits with a fine-toothed comb. Delousing of this type is a four to six hour operation.

Body lice can transmit three diseases to humans:

√ **relapsing fever**

√ **trench fever**

√ **epidemic typhus**

Of the three, only typhus has presented a gruesome past. At one time, it was one of the world's most feared and devastating blights, causing widespread pestilence and, as late as the 1920s, was responsible for over three million deaths in post-war Europe. The causative agent, a spirochete (similar to the one that causes syphilis), is passed to the louse while feeding on an infected human. Then, in turn, it is passed in the feces of the louse to enter humans by way of the louses' excrement. We inhale it, or rub it into sores or into the fluid around the eyes and mouth. But we're not the only ones to suffer. Typhus also kills the louse. So, from the view of the louse, it's the human who is the curse—the shameless tormentor. Typhus, by the way, is no longer a major threat to human populations. But to paraphrase Dr. Zinsser: As long as lice prefer the taste of our blood, the threat of typhus remains.

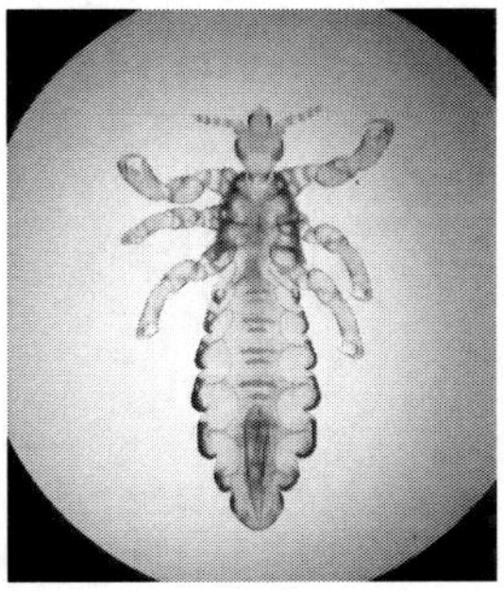

Topic 17

A Campus Challenge

One April morning as I strolled across campus I found myself in the middle of an embittered battle. About 10 yards in front of me stood a male Canada goose. Nearby was his mate. In front of him was another gander equal in size and, at least to my eyes, equal in appearance. The mated male, upon spying the interloper, immediately charged. The meaning of the attack was inescapable. He was determined to dislodge the intruder from his domain...and that included, above all else, his female. The foray was made *en grande tenue*—in full battle regalia—goose regalia, that is. With his head lowered, beak partially agape, wings spread, and feathers upright, he attacked. All the while he uttered his best rendition of a calvary charge...delivered, of course, goose style. The best I can describe this outcry was as a bellowing, drawn out, raucous, wonk...w-o-n-k, getting louder and more frenzied as he faced the trespasser.

But the intruder was not to be outdone. He also Charged—his feathers erect, his head lowered, his wings also held aloft. He bellowed the same battle cry as he lunged forward. They stood face-to-face, not three feet apart, each proclaiming his own grievance. This was obviously something never encountered by our goose. An interloper who did not retreat. But he was determined. He charged again. His rival did likewise. They again rushed each other with equal intensity. They scolded each other with equal fervor. They signaled, postured, strutted, lunged with lusty vigor, each matching the other. The air was filled with their energy. This was true warfare, goose style. A duel in the shadow of ivied halls.

For several minutes the battle raged. Charge after charge was met with amazingly similar countercharges. Each posture, each move, each eruption was matched perfectly by the challenger. Try as he might, our defender couldn't dislodge the invader; he couldn't even induce a ruffled feather. After an exhausting series of assaults and counter assaults, each finally gave up. Our male turned to his mate, gave a few recognition waggles (geese pairs shake their heads at each other to show their white cheek flags as a sign of recognition) and the two sauntered off. To me, as they disappeared over the hill, our gander's strut seemed less confident, his shoulders were slightly drooped, his "honk" less menacing, his eyes less spirited. In his mind he was not the hero he wanted to be. He had failed in this age-old contest of out-bluffing one's foe.

I turned to view the gate-crasher, but he, too, had mysteriously vanished. And no wonder. He didn't exist. He wasn't real. He was but a mirror image. You see, this all took place at the entrance to the science building, in front of the glass doors. The male was charging his own reflection. Each act was met with an equal act—his own. But he didn't realize it. He never figured out that he was in mortal combat with himself.

IT'S A FACT

THE MAJORITY OF ANIMALS BEHAVE INSTINCTIVELY, NOT RATIONALLY. THEY COURT, REPRODUCE, AND DEFEND TERRITORIES BUT WITHOUT A CLUE OF WHAT THEY ARE DOING OR WHY THEY ARE DOING IT. IT'S ONE OF THE STRANGE CONTRADICTIONS OF ANIMALS: THEY FLOURISH WITHOUT KNOWING HOW OR WHY.

Topic 18

Cold Blooded... Warm Blooded

Though boys throw stones at frogs in sport,
the frogs do not die in sport, but in earnest.
- Bion

Amphibians and reptiles are ectothermic tetrapods; in fact, they are the only ectothermic tetrapods. What on earth is an ectothermic tetrapod? The simplest definition of a tetrapod is that it's a vertebrate animal with four limbs, two in front and two in back. There are only four groups of animals that fit this description: amphibians, reptiles, birds, and mammals. An *ectothermic* animal, on the other hand, is one that is unable to control its own internal body temperature. It relies, instead, upon ambient (environmental) heat for **thermoregulation**. It cannot manufacture or maintain its own temperature; it's "cold-blooded." Of the four kinds of tetrapods, only amphibians and reptiles are ectotherms.

So, the binding forces that unite amphibians and reptiles is the presence of four limbs and the inability to control internal body temperature. These two groups of animals are wrapped by the same blanket as it were—blood brothers of sort.

Being cold-blooded has some real advantages. First of all, an ectotherm has a reliable source of body heat: the sun, or in the case of amphibians, water. It uses very little of his own energy to warm up; all it has to do is expose itself to sunlight or situate properly within a column of water. This is a cost-effective way to maintain one's internal heat. The animal expends very little of its own energy reserve to do so. All it has to do is position itself for optimum heat exposure.

Conversely, when external conditions become unfavorable, say it's too cold or too hot outside, an ectotherm just slows down. Its metabolic rate drops, its heart rate drops, respiratory rate declines, and its demand for food dissipates. The animal simply "rides out the storm" until conditions change. This is especially obvious during nighttime; when the temperature drops at night, metabolism likewise drops.

Cold-blooded animals, it seems, are able to coordinate their behavior to external conditions. This is nowhere more apparent than during extended cold periods, such as winter. It is then that ectotherms locate below the frost line and enter prolonged states of inactivity called **hibernation**. As temperatures rise in the spring, they

Figure 18.1 Frogs (upper right), salamanders (lower right), turtles, and alligators (left) are all "cold-blooded," (ectothermic) animals. They often bask in the sun to warm themselves as seen in these three photographs.

Figure 18.2 Sunning can sometimes result in overheating. Here a crocodile dissipates extra heat by panting.

Figure 18.3 In mornings when reptiles are sunning, they are slow and highly vulnerable. Note how this iguana is exposing his body to receive the best rays of the sun.

warm and resume their activity. Some, however, use a different strategy and the result can be amazing. There are frogs of the genus *Rana* that occur in northern temperate zones (e.g., Canada) that are able to survive the coldest of winters not by relocating below the frost line, but by actually remaining exposed to frigid temperatures. They freeze. Their tissue even becomes partially crystallized. In this state of suspended animation, they remain frozen, but alive. This is possible because their blood contains a high level of glucose, which acts very much like "antifreeze" and prevents cellular death. When the temperature warms and ice melts, the frogs, too, amazingly thaw out.

In comparison to warm-blooded animals, ectotherms tend to be highly efficient energy converters. They require less food than a bird or mammal of comparable size. In fact, they use, on average, about *3% of the food needed by a similar-sized mammal.* Birds and mammals expend about 98% of their energy intake just to keep warm. That leaves very little for other functions, such as growth and repair. You will never see a bird or mammal regenerate a tail, for example! Amphibians and reptiles, on the other hand, relying on the sun for warmth, convert about 50% of their energy into new tissue. They are simply more efficient in their use of food. As a result, ectotherms can survive in situations of low food supply unsuitable to birds and mammals.

However, being an ectotherm is not all positive; it has some drawbacks as well. For instance, their geographic distribution is severely limited because of ectothermy. While they may love warm climates, only a few species can survive well in cold ones. That's why you find a superabundance of frogs and snakes in the tropics but virtually none in the tundra. There are very few snakes or toads in Alaska. They are rare on mountain tops. If you want to collect frogs or snakes, you go to tropical rain forest; if you want to avoid them, go to Siberia.

The very fact that cool weather causes amphibians and reptiles to slow down also makes them vulnerable. If you are on a lizard collecting excursion, the best time to catch one is in the early morning while they are "sunning." Later in the day, they are just too fast or too alert to capture. So, during cool periods, such as at night, early morning, and evening, they are pretty much defenseless.

There's another side to the vulnerability story. Not only are they exposed to predators during cool temperatures, but, because they are slower, they are less likely to catch their own food. There's a real balancing act between lowered metabolism and actual starvation. Being ectothermic may offer some real advantages, but it has some serious liabilities as well.

Staying Warm...Keeping Cool

It's a cool April evening some years ago on the Anza Borrego Desert in southern California. The class I'm teaching on Vertebrate Natural History is spending a day observing wild animals. Our goal is to record any and all vertebrates seen during one 24-hour period. We arrive around noon to set up camp and I issue my usual instructions ("Always use the buddy system"..."Don't eat the cactus flowers"..."Don't lick the lizards"...and "Under no circumstances are you to pick up, molest, or handle any rattlesnake").

Exactly one day later, following a sleepless night and many hours in the field, we met to compile our data. The results reveal two worlds in one: a daytime fauna that differs significantly from that of the nighttime. During the daylight hours, although mammals are recorded, they are never plentiful. It's the reptiles that rule the daytime. For instance, on this particular visit we recorded over a dozen species of lizards, including such forms as horned "toads," chuckwallas, iguanas, skinks, whiptails, and geckos; eight different snakes (rattlers, rosy boa, shovel-nosed, lyre, gopher, and king-snakes); and even a desert tortoise.

However, the picture changed abruptly at dusk. As the temperature dropped, so, likewise, did the number of reptiles. The cooler it became the fewer snakes and lizards we encountered. In fact, the only reliable place for reptiles during a cold night was on the asphalt roadways where heat from the sun still lingered. As reptile numbers declined, they were replaced by kangaroo rats, kit

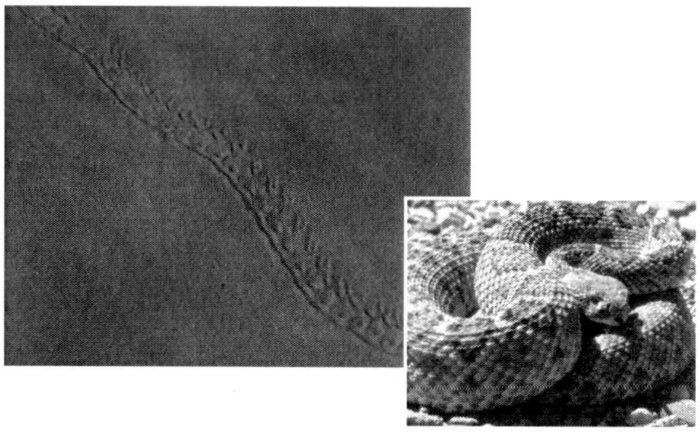

Figure 18.4 Lizard tracks in the sand (left) and a diamondback rattlesnake, a common desert resident.

foxes, woodrats, pocket mice, cactus mice, and bats. It seemed to us that, except for owls, the night was governed by mammals.

On the surface the picture reflected a rather simple dichotomy: days were given to reptiles, nights to mammals. But closer scrutiny revealed a somewhat different, and more complex, pattern. In reality, all vertebrates (reptiles, birds, and mammals) were active during the late morning and early evening hours but few were recorded during the hottest part of the day. At night, when temperatures were the coolest, reptiles were surprisingly scarce, but mammals were abundant.

Endothermy - the internal furnace

Daily patterns of vertebrate animals seem to differ considerably within desert environments. To understand why that is so we need to look to their physiology, more specifically, to the ability of an animal to regulate its internal body temperature. We have already seen that reptiles are ectotherms. Their internal temperature depends upon environmental conditions. Since metabolism (cellular function) is more efficient at warmer temperatures, it stands to reason that ectotherms will be more active during the warmer part of the day.

So how then do we explain the relative inactivity of all animals in the afternoon hours, when it's really warm…hot, actually? The answer, again, is found in their physiological makeup. It's true that metabolism is more effective at warmer temperatures, but only to a point. Danger lurks if the body gets too warm. As internal temperature rises above a critical level, cells begin to dysfunction, metabolism breaks down, the body dehydrates, and death is a real possibility. So, when it's really hot on the desert, reptiles seek shelter. But so do birds and mammals. Thus, we recorded few overall animals out and about on hot afternoons. The same factors dictate reptilian behavior at nighttime. On cool nights reptilian metabolism cannot keep pace so it slows down; reptiles become lethargic and inactive, but nocturnal birds and mammals don't. Why is that? Again, one word provides the answer: **endothermy**. Birds and mammals are endothermic (warm-blooded). They can adjust and maintain their internal temperature irrespective of outside conditions. When it's warm or cool they are active, when it's hot they can still be active (but with vigilance), and when it's really cold they can be active then, too, but again, cautiously. Mammals, using their keen sense of smell have taken largely to the nighttime while birds, being primarily visual animals, have exploited the daytime.

Birds and mammals are the only animals able to self-regulate their internal temperature; they are the only endothermic animals on earth. If that's the case, there must be some advantage, some value to endothermy for these animals. The nature of that advantage is clear from the above paragraph: **activity**. *An endothermic animal is an active one.* They are not shackled to the whims of climate; their metabolism is not linked to a vicarious environment. Because they have an internal furnace that keeps them warm under virtually any circumstance, they are free to be active regardless of outside temperature. As a result, they can invade habitats not open to reptiles. Birds and mammals make up virtually all of the Alaskan vertebrate fauna. They are dominate in Siberia, Canada, Patagonia, the Arctic, and the Antarctic. They are the featured vertebrates on mountain tops, on the taiga, and on the tundra.

Endothermy, then, seems to make an animal more independent, more prosperous. That's certainly true, but it's achieved at a cost. There's a price to pay for our constant level of activity, and *the price is energy.* It takes a lot of energy to regulate our temperature and to remain active. When it gets cold outside reptiles slow down, their metabolism moderates, energy demand drops, and they cease eating. Birds and mammals, on the other hand, to remain active must keep stoking that metabolic furnace, and food is the needed fuel. As much as 98% of the energy budget of an endotherm goes to temperature maintenance. Thus, birds and mammals must keep eating. Reptiles can go days, weeks, even months without a good meal; non-hibernating birds and mammals are dead in a few days. So it's a trade-off. Remaining active is an evolutionary advantage, but endotherms must consume tremendous amounts of food to reap that prosperity.

Special Adaptations

Endothermy very likely evolved in mammals as they adapted to an insectivorous, nocturnal lifestyle not available to ectothermic reptiles. This is because early mammals probably maintained their body temperature close to the ambient nocturnal range. Later, they most likely became more active during the daytime (diurnal) and maintained body temperature at a higher level. This enabled them to eventually survive in a wide variety of habitats, including dry, hot deserts, tropical rain forests, and temperate zones, as well as in extremely cold arctic environments. This led to a more active and agile animal with a number of specific adaptations.

(1) Limbs are no longer splayed out to the sides as in many reptiles but, instead, became rotated beneath the body for more efficient balance and locomotion. Mammals are easily the fastest and most nimble of all land animals.

(2) A more active lifestyle requires a more efficient supply of oxygen. Lung surface area, thus, is enlarged and a muscular **diaphragm** is used to alter pressure inside of the thorax, which, in turn helps to ventilate the lungs while breathing.

(3) The circulatory system improved to better deliver oxygen to the cells. The four-chambered heart is perfected to supply a constant high level of blood pressure, which, in turn, assures timely blood supply through the thousands of miles of capillaries to the trillions of cells.

(4) Likewise, the digestive system underwent improvement through the addition of fingerlike extensions called villi and microvilli. These carpet the small intestine to increase surface area for digestion. Certain ruminants (e.g., cattle, sheep, bison, goats) have developed a multi-chambered stomach that permits them to re-chew their food and to help with the digestion of the massive amount of plant material consumed.

(5) The teeth of mammals and the chewing apparatus have also undergone extensive modification.

(6) Kidneys are designed to remove an increased volume of urea, which is stored in an expandable urinary bladder. The kidneys also are well adapted to regulate the body's water content.

(7) Finally, mammals possess three important skin glands, each with a specific purpose:

a. sweat glands help to regulate body temperature.

b. sebaceous glands secrete sebum, an oily substance that serves to waterproof and soften the skin.

c. apocrine glands are modified sweat glands that secrete a sexual attractant, much like a pheromone.

The Predator-Prey System

In natures' great game there is no right or wrong, no book of rules...competition is the arena and survival is the only trophy.

The Great Game

Predation, in its broadest sense, is defined as the consumption of one living organism by another. Although this sweeping definition includes herbivory, the consuming of plants by animals, we will adopt the more common interpretation to include only carnivory...one animal, the **predator**, eating another, the **prey**.

Ecologically, the predator-prey system is a classic example of energy transfer. A predator uses the prey as food, an energy source. But the interplay is far from just a simple transfer of energy. Indeed, close examination reveals a complex arena where the two participants play a dangerous game of survival. The prey is ever on the defensive, trying to outwit, outrun, outmaneuver, and outbluff the predator. Not to be outdone, the predator tries just as hard to outsmart, outpace, or overpower the prey.

Each encounter can result in sudden death for one and a meal for the other. Since there can only be one victor, each is deadly serious as they grapple together in this ongoing struggle. The prey develops elaborate defenses to prevent catastrophe and the predator evolves equally intricate devices to puncture those defenses. Through this age-old interplay they each respond genetically to the devices of the other. The prey evolves in reaction to the predator, and the predator counter-evolves as it adapts to the prey. Like living boomerangs, they bounce back and forth in a contest of genetic one-upmanship.

In evolution's brazen game, a slender advantage seems to be on the side of the prey. After all, the selective pressure is greater for them. If their defenses are inadequate, they pay the ultimate price. Mother Nature is quick to eliminate the slow, the dimwitted, the careless, the straggler; but she is just as quick to favor the fast, the smart, the careful. Evolution can be kind toward the fit.

Predators, on the other hand, usually have more than one prey species from which to choose, so pressure on them is not so urgent. As a result, the pursued seems to keep slightly ahead of the pursuer in this genetic seesaw. Even so, both lurch ahead to gain an advantage through a rich panoply of behaviors, weapons, tools, and schemes.

A Matter of Numbers

One of the most distinguishing features of the predator-prey system is that prey always outnumber predators. It's a simple matter of numbers. There has to be more flies than spiders, more shrimp than squid, more squid than sharks, more mice than barn owls, more snowshoe hares than lynx. There must be more deer than wolves and more gazelles than leopards. The very survival of the predator depends upon a plentiful and continuous supply of prey animals. In ecology, this is represented as an "energy pyramid" (below). At the pyramid's base are the vast autotrophs (the producers = phytoplankton or green plants); above them are the heterotrophs (the

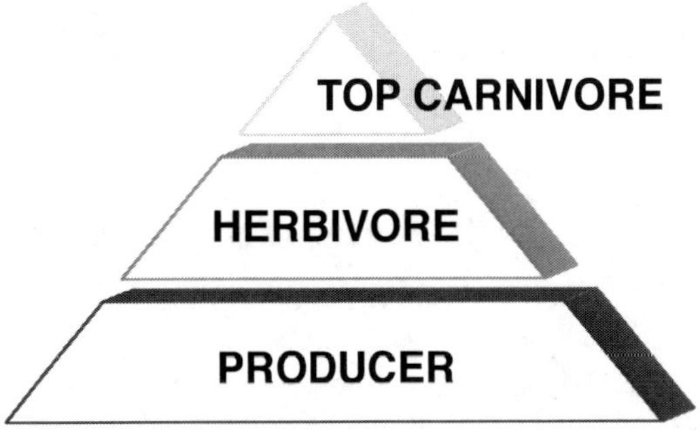

Figure 19.1 Population size can be depicted as an energy pyramid. The lowest levels are represented by autotrophs (producers), primarily phytoplankton and green plants. These producers far outnumber those above, and the number of top carnivores is the smallest in any ecosystem.

consumers = herbivores, omnivores, and carnivores). At the very apex of the pyramid are the top carnivores. This simple sequence is typical of energy as it moves through various trophic levels within any ecosystem. Each level outnumbers the level above it. The producers outnumber herbivores, who outnumber carnivores. Numbers decrease as you travel up the sloping pyramid with the top carnivores having the smallest population sizes of all.

How much a prey population must outnumber the predator is a product of both *size* and *selectivity.* Let's suppose that a relatively large predator such as a coyote has insects, mice, and rabbits available within its feeding range. Let's also suppose that the coyote must select only one type of prey. You can easily see that it would take more insects than mice to support a single coyote. Likewise, there would need to be more mice than rabbits. Thus, to the predator, the number of prey needed for survival depends largely on prey size. If, on the other hand, the coyote is less selective and feeds on insects, mice, and rabbits, then numbers of each prey population can be more flexible. The predator can just sample a few prey from each population. Or if the level of one population decreases, the predator can shift its efforts to the other two. But if the predator relies only on one prey species, then that population must be relatively large.

On the plains of South Africa there is a predator- prey system that well illustrates the relationship between predator survival, prey density, and prey selectivity. The aardwolf (*Proteles cristalus*) is a member of the hyena family that, unlike its ferocious cousins who prey or scavenge on large game, feeds almost entirely on a single species of insect, a termite. The snouted harvester termite (*Trinervitermes* spp) is small, even in termite terms. It lives in mounds and comes to the surface at night to forage. The habits and anatomy of the aardwolf are linked to the activity of their tiny prey. Each night the aardwolf searches out long columns of foraging termites, then laps them voraciously with a long, pointed, sticky, and mobile tongue. Imagine, a predator nearly the size of a German shepherd dog feeding entirely on 1/4"- long termites. A single meal would easily consist of thousands and thousands of termites; indeed it is estimated that over 200,000 are eaten each evening by each aardwolf. The number of snouted termites must be astronomical to support a prosperous aardwolf population.

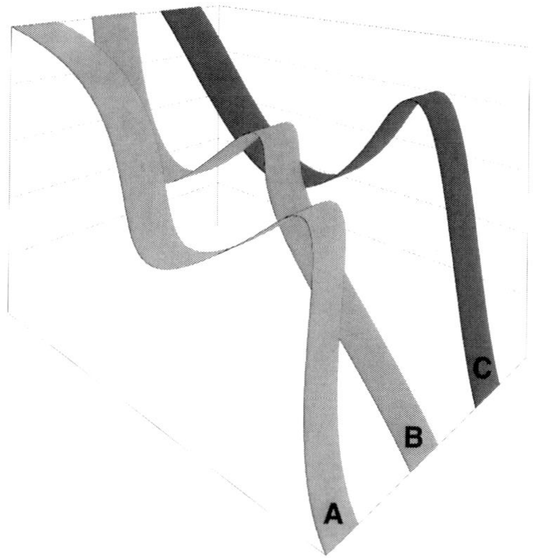

Figure 19.2 Fluctuations in animal populations often follow a predictable and well-established pattern as shown here. Curve **A** represents a food source for a population of herbivores (curve **B**). As food levels dip and then decline precipitously, so does the herbivore population. This is soon followed by a similar drop in any predator population dependent on the herbivores as indicated by curve **C**. Because of the decline in the food supply (A), both population B and C also declined.

When Prey Numbers Crash

Prey numbers, as important as they are, rarely remain constant. Instead, they tend to fluctuate—at times upward, at other times downward. Under certain circumstances, the fluctuation is remarkable. When conditions remain favorable, a prey population may increase generation after generation until incredible numbers are reached. But conditions change, and when they do...watch out! If food becomes scarce, if drought occurs, if temperature shifts, then an otherwise growing population may begin to slide. The decline may be slow and gradual or fast and precipitous. At times, circumstances may be so severe that drastic crashes occur. Since predators depend upon a plentiful supply of prey, they too inevitably suffer. The whole stack of cards may come crashing down as shown above

Crashes can have serious consequences for predators since they rely so heavily on prey numbers for their survival. If the predator doesn't somehow respond, its population may become so deteriorated that recovery to previous status is impossible. Or worse yet it may suffer extinction. So how then does the predator respond to such crashes? It seems they do have a few options available.

(1) When prey numbers plunge a predator may respond by *increasing its foraging range.* In fact, normal patrolling areas may be expanded significantly. For example, the usual foraging range of snowy owls (*Nyctea scandiaca*) is the Arctic tundra, reaching south to the Canadian border. But during times of prey shortages, especially in lemming populations, snowy owls may migrate far south. Under such circumstances, they have been spotted feeding on the southern shore of Lake Michigan and further south.

IT'S A FACT

A COMMON ASSUMPTION IS THAT PREDATORS, BY THEIR VERY NATURE, ARE THEMSELVES A MAJOR CAUSE OF PREY CRASHES. WHILE THAT IS A POPULAR IDEA AND MAY AT TIMES OCCUR, IT SEEMS TO HAVE LITTLE FACTUAL BASIS. FOR PREDATORS TO INDUCE SEVERE POPULATION DECLINES, THEY WOULD HAVE TO SUPPRESS OR SOMEHOW REGULATE THE PREY'S REPRODUCTIVE ABILITY. IN OTHER WORDS, THE PREDATOR MUST REMOVE A LARGE PERCENTAGE OF REPRODUCTIVE PREY TO CAUSE SUCH CRASHES.

HOWEVER, VIRTUALLY ALL STUDIES INDICATE THAT PREDATORS DO NOT SELECT MATURE, HEALTHY INDIVIDUALS, BUT RATHER CONCENTRATE ON THE WEAK, YOUNG, OR UNHEALTHY...WHAT ECOLOGISTS TERM THE "DOOMED SURPLUS." TO CONCENTRATE THEIR EFFORTS ON HEALTHY PREY IS JUST TOO COSTLY; IT TAKES TOO MUCH TIME AND TOO MUCH EFFORT. ACTUAL PREDATION, IN WHICH NON-REPRODUCTIVE INDIVIDUALS ARE REMOVED, HAS LITTLE EFFECT ON OVERALL PREY NUMBERS. SO IT SEEMS THAT CRASHES ARE RARELY ATTRIBUTED TO THE ACTION OF NONHUMAN PREDATORS. THE CAUSE OF CRASHES IS INSTEAD ALMOST ALWAYS DUE TO NONPREDATORY FORCES SUCH AS SCARCITY OF FOOD, CHANGE IN CLIMATE, DROUGHT, DISEASE, AND/OR HUMAN INTERVENTION.

(2) Another option during bleak times is for a predator *to expand its food selection.* It is well known that predators, like all animals, have preferred prey species: snowy owls prefer lemmings, wolves prefer deer, foxes have a liking for mice. But during a time of crises, when that favored source is scarce, the predator may have no choice but to feed on a wider range of less preferred food. If mice populations decline, foxes may eat berries. When rabbit populations fall, coyotes start looking favorably on insects. In fact, adopting a more general feeding strategy is probably the most common reaction to a slump in preferred prey numbers.

(3) One short-term solution to food shortages is also an extreme one. But then, extreme circumstances sometimes call for extreme solutions. When facing starvation, predators may turn on each other. **Cannibalism**, as harsh as it sounds, may actually have some positive effects. First, it provides an alternate food source for an animal facing starvation. Second, it reduces competition for what few prey might be available. And, third, cannibalism often weeds out the weak, sick, or old within the predator population itself. Thus, because of cannibalism, when prey species rebound the predator population may actually consist of younger, more vigorous individuals. In other words, cannibalism can serve to cull out the less fit among one's own species. However, over an extended period of time cannibalism is a maladaptive trait, especially if the cannibals become too aggressive or begin to feed on their own offspring. For that reason, long-term cannibalism is considered an evolutionary liability and is rarely seen except in cases of chronic stress.

(4) One other way that predators may respond to severe states of malnutrition is by *altering their reproductive patterns.* Food shortages can ultimately reduce a female's ability to become impregnated or may limit the number of offspring she produces. Those young that are issued invariably face a bleak future and suffer a high rate of mortality. There is also ample evidence that under severe stress the male or female parent will even cannibalize their own young. The effect of all of this is, of course, to limit the size of a predator population. As a result, there are fewer of one's own kind to compete for the available food, which, in turn, enhances the survival chances of those that remain.

Figure 19.3 When preferred prey numbers decline, a predator may respond by feeding on nonpreferred prey. A coyote, for example, may begin to eat insects instead of mice.

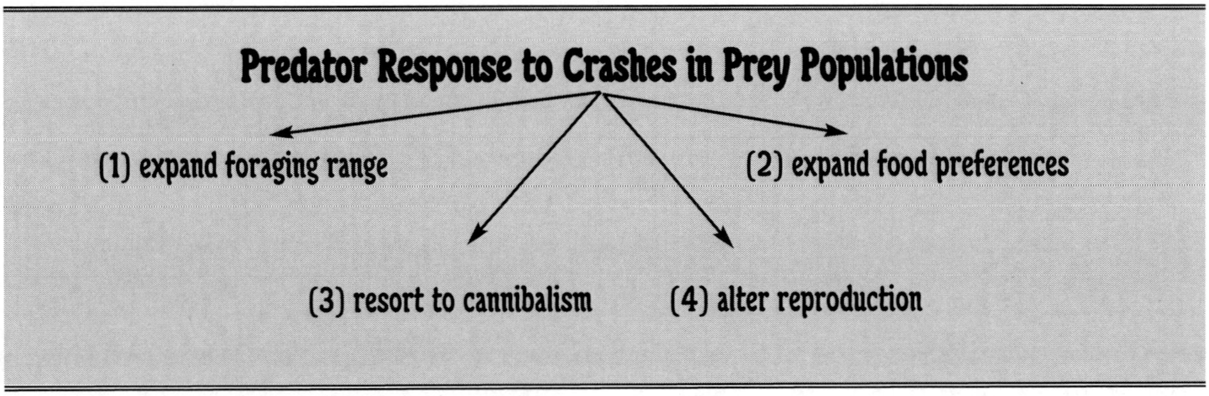

Topic 20

Co-Evolution

Figure 20.1 A deer stands alert sensing possible danger.

Both participants in the predator-prey interaction are affected by it; neither remains neutral. They both, in fact, show a number of special adaptations that promote survival. As selection pressure increases on the prey, it improves overall means of defense. It becomes faster, craftier, or more deceptive. Likewise, the predator improves its offensive course of action. Predators may become larger, stronger, faster, or more aggressive. Whatever the particular response, there is an evolutionary kinship between the two populations, a mutual give-and-take resulting in what is termed **co-evolution**.

The interesting feature about co-evolution is its reciprocal nature. As the prey improves its defenses, it causes the predator to respond likewise. Then as the predator counters by enhancing its offensive arsenal, it, in turn, elicits even better responses on the part of the prey, and so on. In essence, *co-evolution is an interplay of adaptation and counter-adaptation to ever-changing pressures brought on one population by another.* A never ending ebb-and-flow in adaptations between the two populations occurs.

An excellent example of reciprocal evolution is in the development of sensory structures. Both prey and predators must be acutely aware of what happens around them. The survival of each depends upon it. The hunted must pay meticulous attention along a road fraught with danger where misfortune waits at every turn. The hunter, too, sits on a precarious perch since meals are few and far between for the dull-witted predator. Thus, the tendency is for both to hone their awareness skills through sharp vision, a keen sense of smell, acute hearing, a delicate sense of touch, and overall vigilance. Back and forth, each continually adapts to the other through nature's ever-changing landscape

Figure 20.2 Predators, too, have a keen sense of awareness as indicated by the cougar (above) and shark.

Predator Adaptations

Successful predation requires several distinct and recognizable steps. To thrive, a predator must find, capture, subdue, and eat its prey. If there is a disruption in any of those steps, the predation attempt fails and the hunter goes hungry. Predators have, therefore, evolved a number of features and strategies to maximize their success.

Hunting Tactics

Predators have developed three distinct hunting patterns to locate and capture their prey: stalking, pursuit, and ambush. Each tactic has its specific advantages and disadvantages. For example, **ambush** requires that a predator spend much of its time lying in wait for a prey to appear. This is a risky practice since a suitable prey animal may not materialize for some time. However, since this is a passive process, the predator expends very little energy in doing it. Thus, the lie-in-wait mode of hunting invests considerable time, has a relatively low success rate, but involves minimal energy expenditure. The cost in time is balanced by the energy conserved. The ambush style of hunting is characteristic of many insects, spiders, frogs, toads, snakes, owls, and alligators as shown in Fig 21.1.

At the other extreme is the **active pursuit**. Here the predator spends very little time in searching and virtually no time in waiting but exerts considerable energy in the capture itself. Therefore, the prey must be relatively large to provide enough food to offset the amount of energy spent in pursuit. In this tactic, time investment is low but the energy spent is high. Active pursuers include many fish, hawks, some sharks, a few of the big cats, wolves, hyenas, and hunting dogs.

The final hunting category seen with predators is the **stalk**. Here the hunter can pick from two strategies. First, the predator may spot a prey animal, slowly follow it, then explode in a fierce attack. Such is the practice of leopards, cheetahs, lions, and many other deliberate

Figure 21.1 The three hunting strategies of predators: **A**, the lie-in-wait tactic of the ambush is shown by this alligator; **B**, a herd of zebra are actively pursued by African hunting dogs; and **C**, a great blue heron slowly stalks fish, crustaceans, or other prey species. Wading birds use their shadow to startle prey which they quickly spear.

predators. A second stalking approach is to gradually explore or canvass a given area until a likely prey is uncovered. This is again followed by a frenzied strike. The roving stalk is typical of herons, bitterns, some raptors, and certain predacious fish. The stalk, whether trailing a victim or locating one randomly, takes considerable search time but involves little pursuit. As a result, moderately sized or even small prey can be taken.

A lone hunter must be careful not to tackle dangerous prey or those considerably larger than itself. However, predators do, in fact, often take prey that fall into this category. One highly successful way to do this is to *hunt cooperatively in packs.* Several predators working together can tire and subdue prey that a single predator could never handle. Certain fish, most notably the piranhas, wolves, hunting dogs, hyenas, jackals, African lions, and sharks often resort to group hunting.

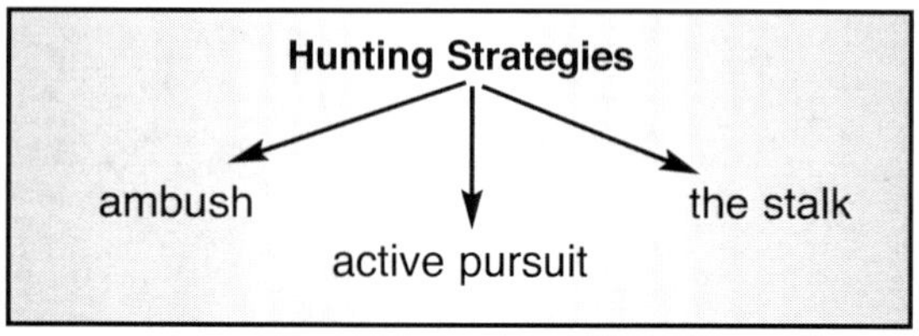

Deception

~ an enterprising con game ~

Many predators owe their success to trickery. An unseen or unrecognized predator is more often than not a successful one. Deception just makes the hunt, whether it's an ambush or a stalk, so much easier. A prey that doesn't spot the predator until that last rush will most likely suffer a sudden death. It is not surprising then that predators have invested much of their evolutionary energies into subterfuge. Some have concentrated on developing confusing colors or patterns, while others have opted for stripes, or spots, or background blending. Still others place success in the form of disruptive coloration, which breaks up the body's outline. Whatever the chosen route, concealment is the goal. The more invisible the hunter, the better the hunt is likely to go.

Some predators even resemble inanimate objects. There are praying mantids, for example, that mimic orchids upon which their prey feed. An unsuspecting moth, butterfly, or other insect doesn't stand a chance against this well-concealed, formidable predator. Likewise, a species of ambush bug matches the yellow florets of the goldenrod plant to surprise any careless passerby. Stonefish and scorpion fish are highly poisonous animals that blend remarkably with the rocks, sponges, and debris of the ocean floor. Unwary prey that swim too close are suddenly engulfed in the vacuum-like mouth of these voracious predators. In a similar manner, alligators and crocodiles have the appearance of floating logs, which enables them to drift unnoticed to within striking distance.

An even more incredible form of deception is for a predator to resemble the prey itself. This allows the hunter to mingle among unsuspecting quarry without detection. An example is the robber fly (*Mallophora bomboides*) that mimics bumblebees with amazing accuracy. In fact, the imposter blends so well among feeding bees that it can intermix unnoticed, and attack without warning.

A similar ploy is used by certain fireflies. As members of the order Coleoptera, fireflies are bioluminescent beetles that use species-specific flashes to signal each other. The females often lie in the grass and use their amorous flashes to lure males. The two meet and mating occurs. However, there is a predacious species that can mimic another species' code. She, too, lies in the grass to lure males, not to a romantic interlude, but to a certain death.

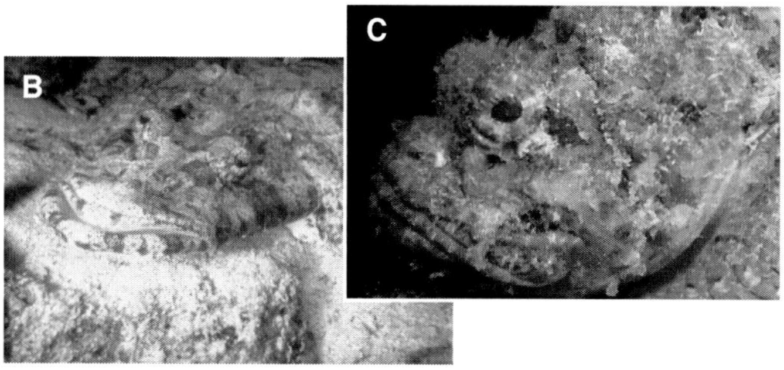

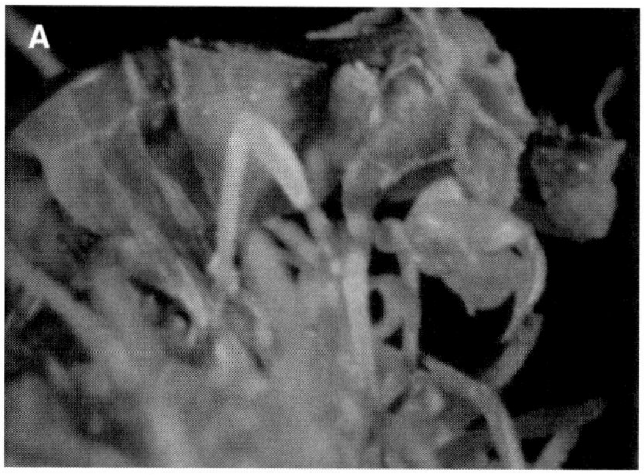

Figure 21.2 Camouflage can greatly assist an ambushing predator. For instance, plate **A** shows an ambush bug that blends perfectly with its background. Plates **B** and **C** display the tremendous camouflage used by predacious fish to match their surroundings while awaiting their unsuspecting prey.

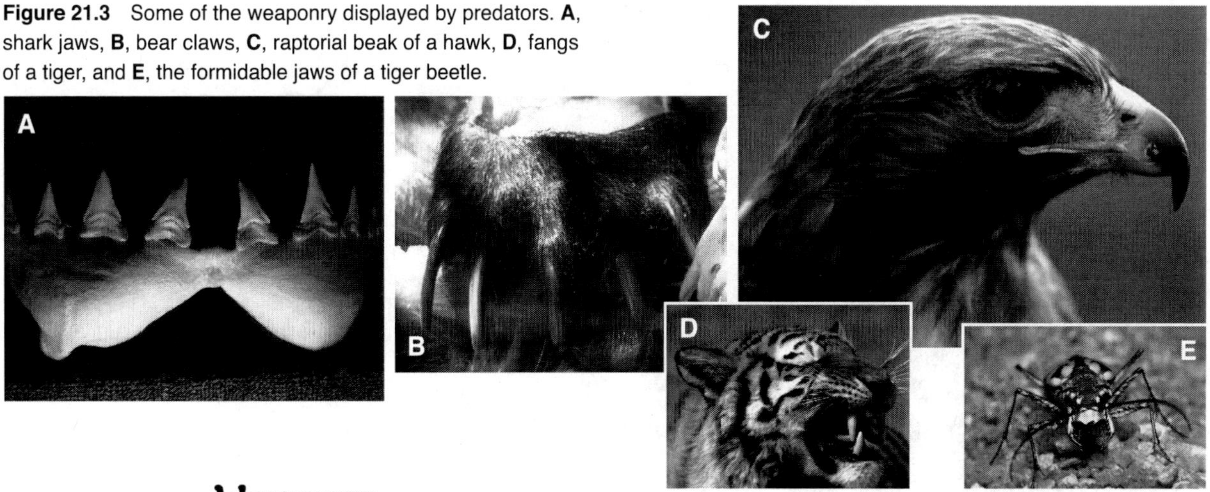

Figure 21.3 Some of the weaponry displayed by predators. **A**, shark jaws, **B**, bear claws, **C**, raptorial beak of a hawk, **D**, fangs of a tiger, and **E**, the formidable jaws of a tiger beetle.

Weaponry

~ the predator's arsenal ~

Locating prey without being detected is but the first step toward a successful hunt. To be victorious the predator must also be able to overpower or somehow subdue the quarry. Thus, it's to the predator's advantage to be larger, faster, and more powerful than its prey. While that's what we would expect, that's not exactly what we find. It *is* true that the vast majority of predators are larger and more powerful than their prey, but they are not always faster. In fact, selective pressure on the prey has often yielded species that are speedier than their pursuers. This is especially true when the confrontation occurs over relatively long distances. So, for the predator it's that initial contact that is of utmost importance. That first burst of speed very often determines the outcome of an encounter. If the predator doesn't overpower the prey in the opening few moments, the contest may go to the hunted. That's exactly why predators have evolved ambush, stealth, and stalking as such important elements in their bag of tricks.

That initial burst of speed may allow the predator to overcome the prey, but ultimate success depends upon having the necessary weaponry to kill. And here, natural selection has outdone itself in providing tools that inflict as much damage as quickly as possible. Elongated fangs, slashing claws, sharp spines, powerful appendages, slicing mandibles, bone-crushing jaws, and powerful toxins are just some of the weapons within the predator's arsenal.

The wide use of *chemical weapons* is another shrewd evolutionary development. From the nematocysts of the cnidarians to the powerful toxin of the blue-ringed octopus, venom is a popular weapon in the predator's armory. Spiders, centipedes, and a host of insects are among the many that can inject venom to overwhelm their prey. And, of course, no predator uses venom more effectively than snakes, over 300 species of which are poisonous.

The bounty of weapons evolved by predators is astonishing. Spiders use sticky webs to entrap their victim, while the angler fish and alligator snapping turtle use specialized structures (fins, tongue) to lure prey within striking distance. Bats and dolphins use echolocation to find their targets, and owls have feathers modified into facial discs and asymmetrical ear openings to locate quarry in virtual darkness. Many fish, especially sharks, have a lateral line system to detect weak electrical fields or subtle vibrations emitted by their prey. Electric eels, electric catfish, and torpedo rays can discharge an electric shock of several hundred volts to stun or kill their victims. The arsenal is truly impressive.

Figure 21.4 Many predators use venom to capture prey. Some, such as the anemones (top) use a passive approach, while others like the copperhead and scorpions strike with sudden death.

Figure 21.5 The facial discs of owls help to channel noises to the ears, which are positioned asymmetrically to better locate the noise source. The "ears" on top of the head are not used in prey location but are for display purposes only.

Topic 22

Prey Adaptations

From our discussion so far, it may seem that the advantage in the predator-prey struggle is well on the side of the predator. It may seem that way, but as we shall soon discover, their advantage, if any, is slight. Keep in mind that both predator and prey have co-evolved within this give-and-take relationship. As the hunter developed better ways to locate, capture, subdue, and devour a prey, so have the hunted developed a countering suite of traits to avoid being located, captured, and eaten. Thus, prey have their stockpile of tricks as well. In fact, with their remarkably diverse array of anti-predator devices to avoid becoming just another item on a predator's menu, they may just be a little ahead in this game.

Reducing Detection

In any contest, the first line of defense is very often the best. This is absolutely true for prey animals. If they can avoid detection, they will not be eaten...it's as simple as that. If an animal is seen, heard, or scented, it is at risk. The need to remain undetected, thus, is foremost.

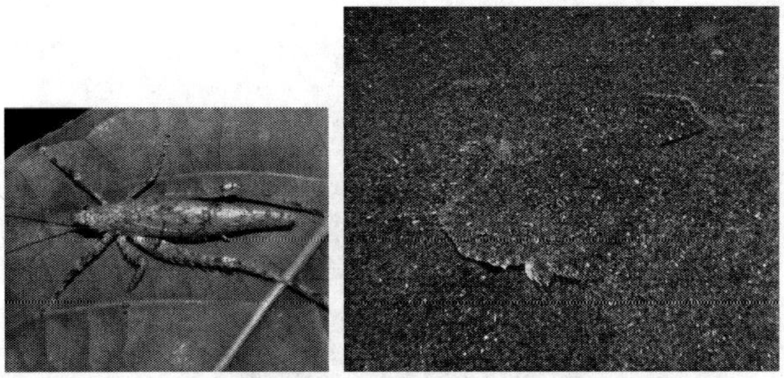

Figure 22.1 Substrate matching is a favorite ploy used by prey to avoid being seen. A fish (upper right) uses color patterns to blend with its surroundings as does the grasshopper above poised on a leaf. The flatfish to the right matches its surroundings so that it's all but invisible.

So, it's not at all surprising that prey, like predators, have devoted a considerable portion of their adaptive effort toward deception. And the art of deception is nowhere more evident than in **camouflage**, the ability of an animal to blend with its surroundings. It enhances their chance of survival and thus increases the likelihood of transmitting these same survival traits to the next generation.

Camouflage is perhaps the most commonly used deceptive device by both predators and prey. The better an animal's coloration or shape allows it to blend with its general surroundings, the more likely it will go unnoticed. Through **substrate matching** a prey animal may blend with local soil, pebbles, dead or living tree trunks, or other features. Fish, reptiles, and many ground-nesting birds show a remarkable ability to melt into the background. To avoid detection, many prey animals have also adopted "**freezing**" as a viable option, especially when face-to-face with a large predator.

An animal's outline poses a particular problem that must be overcome to avoid detection. The problem has to do with sunlight. Overhead light causes shadows to appear beneath an animal and, as a result, a uniformly colored animal actually becomes more conspicuous. So, many animals are darker dorsally than ventrally, which tends to counteract the effect of shadows. **Countershading** is seen in almost all camouflaged species and is especially prominent in aquatic species. When viewed from beneath, a countershaded swimmer blends with bright light beaming from above. That same animal, when viewed from above, blends with the blackness of the underlying water. This is clearly displayed by most game fish, and most predators including sharks (below) and killer whales.

Furthermore, color patterns can be used to distort an animal's shape from a distance. For example, small and large hervibores, from chipmunks to zebras, invariably have stripes or other **disruptive patterns** that break up the body's contour and make it hard for the predator to recognize it as prey.

Figure 22. 2 Prey species display an incredible ability to blend into or match their background. Walking sticks (**A**) are well-known examples of animals that look like inedible features of their surroundings. But so is the leaf hopper that simulates a thorn (**B**), the toad bug that virtually disappears among the pebbles around it (**C**), the gorgonian crab in plate **D**, the lizard hugging the tree trunk (**E**) and the butterfly in **F**, are all able to blend due to their likeness to their surroundings. Another way prey use to conceal themselves is to break up the body's contour through countershading, or with contrasting stripes or spots as seen in the zebra in plate **G**.

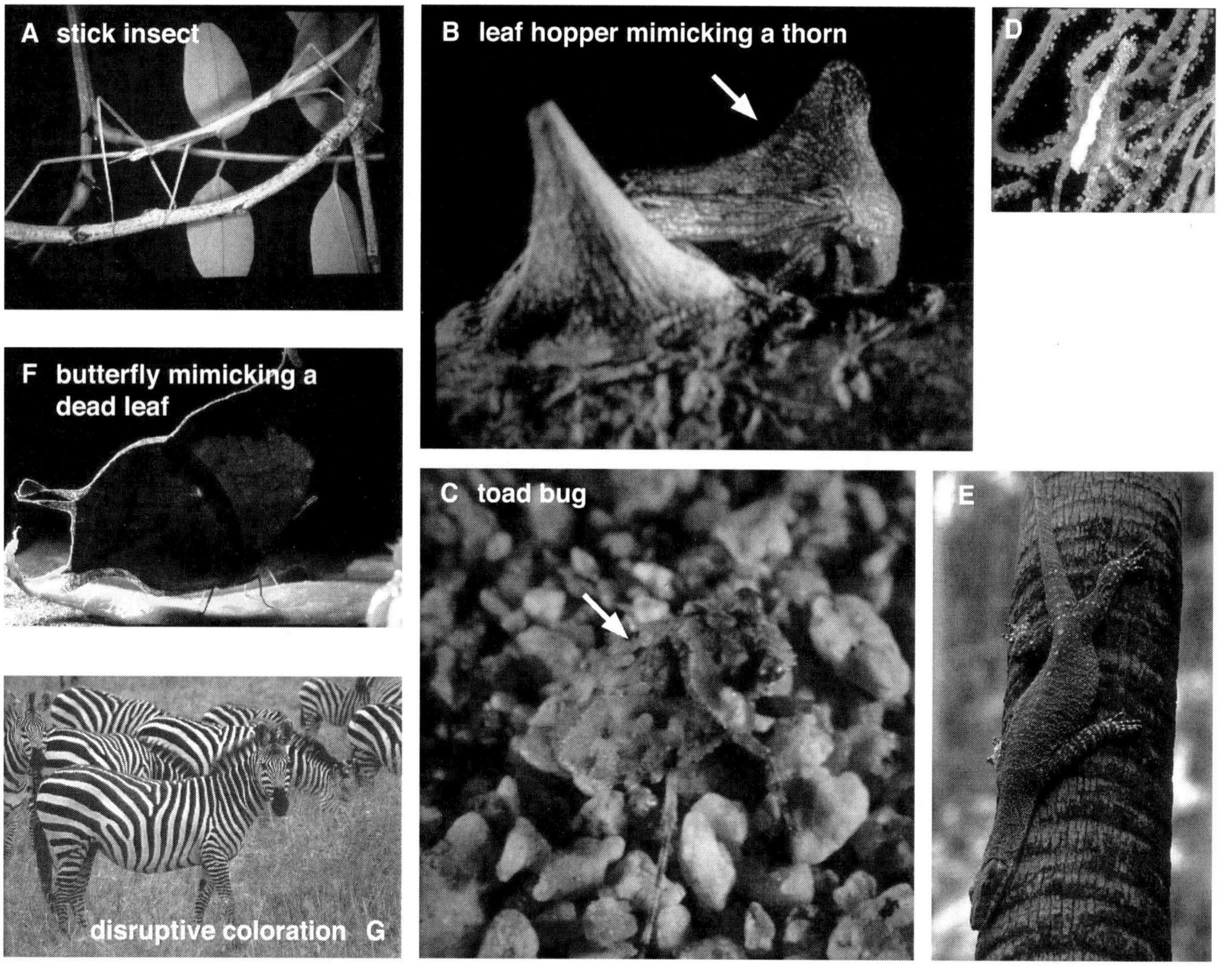

Camouflage isn't always manifested in simply matching an animal's general surroundings. There are many instances of prey resembling specific, inedible environmental features. Leafhoppers mimic thorns, bush-crickets, some fish, and katydids have a remarkable likeness to leaves, caterpillars and walking sticks resemble twigs and sticks. Some prey species, such as caterpillars and other larval insects, may even take on the appearance of bird droppings, lumps of soil, clumps of leaves, and so on. Again, the key is to avoid detection by whatever means possible.

One technique not yet discussed is one of the most obvious (above). Prairie dogs, when catching sight of a predator quickly dart into their protective burrows. Rabbits bolt into thick brush, birds fly off, an octopus slithers into its rocky lair, or a fish ducks out of sight into a thicket of algae, coral, or other sanctuary. Taking refuge may be the most common, and most effective means that prey have of avoiding being seen.

Reducing Capture

Should a prey animal actually be spotted, it must summon any resource available to avoid being captured. Once discovered, a favorite tactic of many prey (as mentioned above) is to take flight or find refuge in a den, burrow, crevice, tree cavity, or other shelter. A different approach, the **aggregation response**, is available to prey that occur in large groups. Many songbirds, for example, flock together to mob a marauding crow, owl, or hawk. When threatened by a pack of wolves, muskox encircle their calves with their massive heads pointed outward. Such group defense has also been observed in meerkats, banded mongooses, capuchin monkeys, and warthogs. In each case, the group bands together to repel a predator.

Another advantage of group living is that each individual in the group can capitalize on a *collective sense of awareness*. Rather than relying on one pair of eyes scanning the horizon, or one pair of ears tuned to the slightest rustle, or one nose sniffing the air, a single prey animal can benefit from hundreds or even thousands of surveying eyes, listening ears, and sniffing noses. It is virtually impossible for predators to spring an ambush in such settings. All they can do is attack quickly with the prospects that a slow-moving animal might be caught in the confusion. However, in reality, predators generally don't waste their energy on such situations.

There is yet one more way that large aggregations of prey can protect the individual. It's called the **selfish herd concept**. In simple terms, the larger the crowd the less the chances that any one individual will be captured. One lone sandpiper feeding on a mudflat may be easy pickings for a hawk or a raiding fox. But, if that same bird is surrounded by a thousand others, its chance of capture is reduced significantly. Even if a hawk or fox is successful in catching a sandpiper, the chance that it is *our* sandpiper is minimal. In other words, the selfish herd is willing to sacrifice one individual for the sake of the pack. From the standpoint of natural selection, we might even assume that it's the slowest, dullest, or least aware that is captured, thus culling out the less fit, allowing those with more adaptive capabilities to survive and reproduce.

As we have seen, camouflage is one of a prey's most important lines of defense. But other aspects of color, beyond simple camouflage, can play a significant role in prey survival as well.

Figure 22.3 At times, prey occur in extremely large groupings so that each member benefits from the collective sense of awareness. Also, the likelihood that any one animal being captured is greatly reduced in what is termed the "selfish herd."

Distractive Coloration

Imagine that a predator, say a blue jay, has spotted a succulent butterfly. Just as the jay strikes, in full expectation of a tasty meal, the butterfly suddenly jerks open its wings to reveal two huge "**eyespots**." The startled jay stops in mid-lunge, turns, and flees. The butterfly has escaped, not because of concealment, but rather due to a precisely placed flash of color. What the blue jay apparently "saw" was a potential confrontation with an even larger animal than itself. This scenario is common in the predatory-prey scheme (see previous page). Eyespots are conspicuously displayed on many butterflies, moths, mantids, fish, amphibians, and snakes, among others. The spots may be large, for intimidation purposes or they may be small, and serve, instead, to deflect the predator's attention away from a vital area.

Figure 22.4 Many prey animals display conspicuous patterns resembling eyespots. These may be large to resemble a much larger animal so that when flashed they serve to repel an attacker or may be small spots scattered over the surface of the prey to simply distract an attacker. Note that in each of the cases shown above, the eyespot is located in a nonvulnerable part of the body so that if an attack does occur the individual may survive.

Distractive coloration, as this phenomenon is sometimes called, is not always manifested in eye spots. Take, for example, the rump of a white-tailed deer. When in danger, a deer will flick the tail sending a white warning signal to other deer in the area. When attacked, it again conspicuously flashes the tail, potentially diverting the attacker toward the rear and away from the head, trunk, or other vital area. Distractive coloration is evident in a large number of animals from insects, to fish, to birds, to mammals.

Figure 22.5 Contrasting patches of color (arrows) is thought to distract a predator toward a less vulnerable area. The color is often concealed until the last moment, then flashed as the prey retreats as seen in the fleeing deer. These flash colors are also used as conspecific signals.

Aposematic Coloration

Warning or aposematic coloration is used by many prey animals to alert predators of some toxic, noxious, or other unpalatable property: a sting, a venomous bite, a putrid smell, or an offensive taste. The idea is for some color combination to *clearly advertise the unpalatability.* Any bold or contrasting bright colors will do, although black or white stripes set against a yellow, red, or orange background is the usual arrangement. Rather than relying upon concealment, as camouflaged prey do, animals with warning coloration do just the opposite. They display themselves conspicuously. When approached by a potential predator, they seemingly flaunt their colors, often turning toward the intruder to show their colors to the best advantage.

Figure 22.6 Aposematic coloration is used by a wide variety of animals to warn of some noxious or dangerous property. Contrasting yellow against a black background is a common pattern as seen in the poisonous flatworm (**A**) and caterpillar (**B**). An exception occurs with birds, which have very few examples of warning coloration. Above is a goldfinch (**C**), also adorned in bright yellow and black. But goldfinches are not noxious; the color is related to courtship behavior and not to warning.

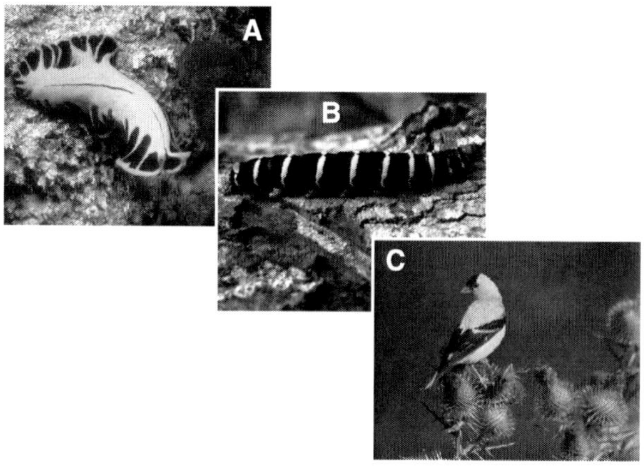

For aposematic coloration to work properly two things are necessary. First, the colors must serve clear notice that the bearer is unpalatable. *The bold color pattern is intended to remind a predator of a previous unpleasant encounter with that species.* Following the initial encounter, the predator will thereafter associate the color with that episode and avoid any animal displaying that color combination. You can now see why brightness and contrasting colors are used. Second, the toxicity cannot be lethal to the predator, otherwise the whole mechanism is self-defeating.

Warning coloration is widespread among prey animals but is especially pronounced in cnidarians (sea anemones), molluscans (sea slugs and nudibranchs), insects (hornets, wasps, bees, ants, beetles, butterflies, caterpillars), fish (lionfish, zebrafishes), frogs (poison dart frogs, tree frogs), salamanders (tiger salamander, fire salamander), lizards (gila monster), snakes (coral snakes, sea snakes), and even certain mammals (skunks).

You may have noted that birds are missing from the above list; yet, many birds are adorned in brightly contrasting colors such as the goldfinch above. It seems that the ability to fly, and thereby escape predation, has left birds pretty much out of the warning coloration business. Few, if any, birds are aposematically colored simply because birds as a group lack stinging devices, poisons,

Figure 22.7 Orange against black, yellow and black contrasted against a white background, red and black—these are the colors that animals have selected to convey the unmistakable message, "don't mess with me."

or noxious smells. There are only a few records of tropical birds whose flesh is apparently distasteful and who display warning coloration. Bright, conspicuous colors are used by birds primarily in courtship displays, not as warning colors.

Mimicry

~ the brilliant art of deceit ~

It's difficult to know exactly how successful warning (aposematic) coloration is in protecting any prey species. We can assume that the process is fairly effective, but we don't know for sure. However, what we do know is that there's a built-in problem with the mechanism: *it's a one-on-one learning experience.* Each predator must have had at least one prior harmful experience before it can connect the color pattern with the unpleasantry. This means that there must be a large number of distressful encounters for the process to work at all. Nonetheless, the fact that mimicry is used by such a large group of animals is testament to its apparent effectiveness.

But popularity is not the only way to measure the outcome of warning coloration; there's one other line of evidence available. You will recall that in the predator-prey contest deception is the passport to success. Nowhere in nature is fraud more evident than in the phenomenon of mimicry: t*he resemblance of one organism to another (or to an object) for the purpose of deceiving a predator.* It comes in four forms.

(1) In **Batesian mimicry** an edible prey animal (the mimic) assumes the coloration of an inedible or unpalatable one (the model). A predator that has had an unpleasant experience with the model learns to avoid the mimic as well. It's a con game of sorts: the mimic gains from warning coloration even though it has no unpalatable properties itself.

Batesian mimicry is seen in a variety of invertebrate and vertebrate animals. Perhaps the most classic example of Batesian mimicry occurs with two snakes, one poisonous and the other not. The several species of venomous **coral snakes** all have a conspicuous pattern of red, yellow, and black stripes. A pattern that any experienced predator will soon learn to avoid. A similar color pattern occurs a variety of harmless **king snakes**, which thereby are equally well protected from predators. In Batesian mimicry, the mimic gains the same degree of protection as the model even though it's nontoxic.

(2) **Mullerian mimicry** also involves models and mimics but differs in one important respect from Batesian mimicry. Rather than one edible form resembling an inedible

Figure 22.8 One good example of Batesian mimicry is that between the venomous coral snake (left) and the harmless kingsnake (right). The North American coral snakes have a banded pattern in which red abuts yellow (arrows), whereas red lies next to black in the kingsnake.

animal, *both the model and the mimic are toxic.* In this case, each prey population is noxious, and each population displays similar warning coloration; to a predator, they all look somewhat alike. Therefore, the predator only has to sample *one* prey animal from *one* population to learn to avoid *all* populations. Again, Mullerian mimicry differs from Batesian in that all participants are toxic and thus the learning curve for the predator is much reduced. As a result, fewer prey must pay the price of educating predators, a price that all too often results in their own death. Mullerian mimicry is common among insects, especially beetles, ants, bees, wasps, moths, and butterflies.

(3) Some prey animals have a keen ability to **mimic inedible objects**. There are birds that perch in such a way that they resemble a tree knot; many insects mimic fecal droppings; other insects called tree hoppers resemble thorns; stick insects mimic twigs; beetles look like stones; there are fish that resemble leaves or floating debris; squid and octopuses that take on the appearance of algae or corals; and hermit crabs that place sponges and sea anemones on their shell. The fact is, many prey animals mirror their surroundings in such a way as to appear not only unpalatable, but inedible.

(4) Many predators will only eat prey they have actually killed themselves and will avoid an animal already dead. There are a number of prey species that take advantage of that through a process called **akinesis**; they pretend they're dead. Certain spiders and beetles will fall motionless when threatened, as do some chameleons. The grass snake and the hognose snake are two species that roll over on their backs with mouth agape and tongue distended when threatened. But the best actor of all is the North American opossum. When attacked, it not only assumes a deathlike posture, which it can maintain for a considerable time, but it, like the hognose snake, can even emit an odor of decomposing flesh. Most would-be opossum eaters are disgusted by this repugnant display of deception. Pretense of this sort works well for the opossum unless, of course, a hungry scavenger that relishes foul-smelling corpses comes along.

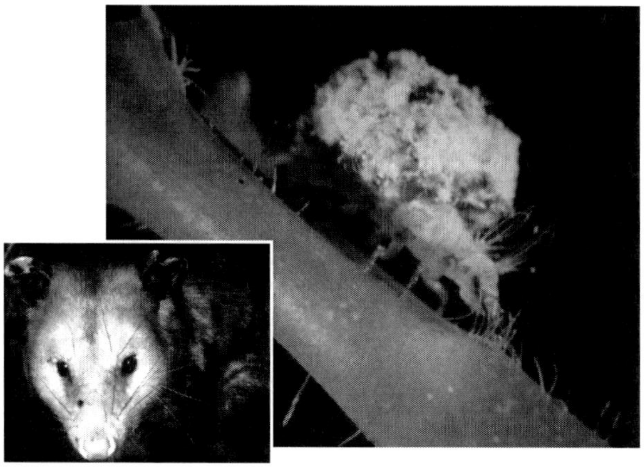

Figure 22.9 The insect above uses its own feces to disguise itself from predators. The inset photo shows the Virginia opossum (*Didelphus marsupialus*), the only marsupial in North America. When harassed, the "possum" is known to realistically feign death, a behavior known as **akinesis**. This is a warm climate species that has moved northward. Note the tattered ears resulting from frost bite.

MIMICRY
(RESEMBLANCE OF ONE SPECIES TO ANOTHER OR TO AN INANIMATE OBJECT)
BATESIAN MIMICRY - RESEMBLANCE OF A NONTOXIC ANIMAL TO A TOXIC ONE,
MULLERIAN MIMICRY - RESEMBLANCE OF SEVERAL TOXIC ANIMALS TO EACH OTHER,
MIMICRY OF INEDIBLE OBJECTS - RESEMBLANCE TO INEDIBLE ENVIRONMENTAL FEATURES SUCH AS TWIGS & LEAFS.
AKINESIS - THE ABILITY OF AN ANIMAL TO PRETEND IT'S DEAD

Preventing Consumption

~ the last resort ~

The final ploy an animal has to avoid predation is aggression...hoping in desperation that a bluff or a counterattack will permit an escape. Some animals puff up to appear large to intimidate an assailant. In fact, when threatened, humans often do likewise. But the champion "puffer" is the porcupine fish (below). It swallows air or water to get bigger and bigger until a harasser gets frustrated and just leaves. Other prey lunge at an attacker and some bare their teeth or emit loud screams in an effort to scare off an assailant.

porcupine fish
(in defensive posture)

But prey *are* inevitably captured. The nature of the game is that predators are successful, at least some of the time. If a prey animal has, indeed, been detected and caught it has but one last defensive strategy and that's to avoid being eaten. There are a variety of ways this can be accomplished.

Chemical defense. The use of toxins is a common prey tactic to gain release. Bees, wasps, hornets, yellowjackets, some species of ants, and a variety of other insects can inflict a painful sting if provoked. Centipedes are capable of rendering a painful bite as are certain aquatic insects. Likewise, many aquatic invertebrates, such as sea anemones, urchins, and sea cucumbers, have toxic armaments to repel predators. Millipedes and centipedes can exude a defensive and poisonous chemical from special glands.

The list of prey relying on chemicals is a long, long one, and some are highly ingenious. For example, squid and octopuses are capable of releasing a numbing ink to confuse and perplex predators. Salamanders, frogs, and toads may ooze a slimy toxic skin secretion—sometimes, as in the tiny poison dart frogs, the secretion is exceedingly poisonous. The bombadier beetle releases an explosive chemical cocktail that burns upon contact. Snakes and a few lizards may inject poisons that can cause human death. But no user of chemicals is more well known than those black and white denizens of the American woodlands: the skunks. When disturbed, they can spray a precisely aimed, nauseating chemical vapor primed to discourage the most determined of predators.

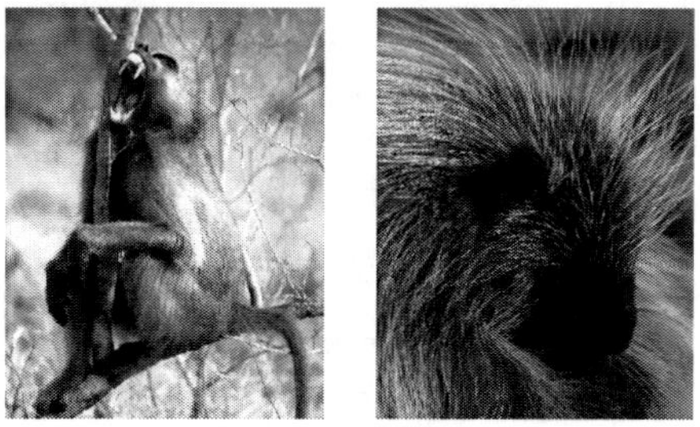

Figure 22.10 Otherwise passive prey can become aggressive when facing certain death. The baboon threatens with large teeth and the porcupine defends itself with sharp quills.

Body armor. An impenetrable covering is another one of a prey's effective safeguards against being eaten. Many animals, when threatened, simply find sanctuary within a shell, casing, carapace, or other armorlike retreat. From amoeba to armadillos, a host of prey animals find refuge in some type of protective covering.

Weaponry. When cornered, many otherwise docile prey will strike back. They may possess teeth, claws, spines, or other devices that can become effective weapons when necessary. Porcupines and hedgehogs have sharp quills that can inflict painful wounds. The bite of a trapped groundhog, skillfully delivered, can discourage a coyote or fox. Large antlers, swinging to and fro, may repel a pack of wolves. Even a claw poking in the right place may be the remedy to an attack. And it's only the bravest of predator who will undertake an attack against a young animal when the mother is nearby. The truth is, predation can be risky business when the life of your victim is at stake.

Autotomy. One of the strangest, and at the same time most reliable, tactics is the ability of some prey to sacrifice a body part when attacked. Self-amputation of this type, called autotomy, is common among a variety of prey species. Many arthropods, such as spiders, daddy longlegs, crustacea (crayfish, lobsters, crabs), and others, will "willingly" relinquish an appendage when in a predator's grasp. Sea stars, for example, readily surrender an arm when the circumstance warrants. Certain lizards will do likewise with their tail. The sacrificed piece may even continue to move or wriggle thereby distracting the predator while the prey makes a hasty escape.

However, for autotomy to be an effective maneuver, the missing piece must be replaced. Thus, this strategy is only used by those prey species able to regenerate lost parts. In most cases, there is a fracture plane where the feature breaks off from the body. The breakage quickly closes to prevent excessive loss of blood or other essential bodily fluids.

Satiation. What happens to those prey species that may lack camouflage? What about those who have no toxic properties, or who cannot mimic, possess inadequate weapons, or don't engage in self-amputation? Are they without means to fend off the horde of predators? The answer is, no. There is yet one more approach open to

Figure 22.11 A drastic maneuver to escape predation is to sacrifice a body part. **Autotomy** works only if regeneration is possible as in the skink (left: tail) and the sea star on the right.

the hunted. A preposterous, yet effective approach. They simply produce an overabundance of offspring so that any marauding predator can eat until it becomes sated. By flooding the market with offspring, they produce more than even an army of hungry predators can possibly consume. Thus, a percentage, albeit a small percentage, of their young will escape.

Many animals, from oysters to insects, from frogs to fish, practice this strategy. Mating horseshoe crabs, for example, deposit millions and millions of eggs on sandy beaches, more than even migrating birds can devour. A world away, sea turtles do likewise (see photo below). In just a few evenings hundreds of female turtles bury batches of plum-sized eggs in sandy crypts. Months later a mass emergence of thousands of tiny turtles litter the beaches as they run a gauntlet to the sea. And even though many are lost to predators, some do invariably make their way to open water and eventual adulthood.

Rabbits, lemmings, and many other herbivores produce large litters. Even foraging ungulates such as wildebeest and caribou have a high annual output. This is all done on the premise that if enough young are issued at the same time, predators can't eliminate them all. Satiation seems to be a wasteful tactic, but it does ensure that some members of the species will carry on into the next generation.

Figure 22.12 A last-resort tactic used by many prey animals is to produce more young than can survive. Thus, predators can eat until sated without eliminating all of the offspring. Sea turtles practice such a strategy through mass egg laying.

Figure 22.13 Some animals, such as these venomous snakes, are efficient predators yet still serve as prey to larger animals.

A Dual Purpose

You have probably already realized that many predators are, at times, themselves, prey. They may prey on smaller animals but serve, in turn, as food for larger ones. Except for top carnivores, hunters are also hunted. Thus, for most animals, adaptations serve a dual purpose; they have both defensive and offensive functions. Camouflage can serve to conceal a predator from its prey as well as from its enemies. A weapon may be used to subdue a victim or to ward off an attacker. Venom can be an awesomely effective way to prepare a meal, or it can be an equally awesome means of self-protection.

The diamondback rattlesnake (*Crotalus atrox*) of the southwestern U. S. is a good example of the duality of adaptations. It is a large, highly venomous, aggressive predator. Well camouflaged, it lies in wait to ambush mice, small squirrels, and other warm-blooded quarry. But its camouflage also provides protection from potential predators (birds of prey, large mammals, other snakes, humans). Concealed within its upper jaw is a pair of large fangs, ready to deliver a lethal dose of venom into a careless rodent, or just as ready to discourage that careless backpacker. It is finely adapted with dual skills as it lives on evolution's cusp between predator and prey.

Topic 23

Final Exams, SATs, and IQs

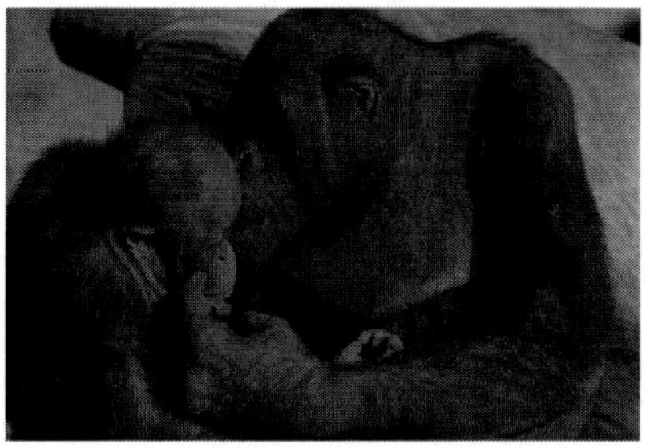

We spend a tremendous amount of time and energy thinking about, and worrying about, how smart we are. Others test us and we test ourselves. How many math tests have you endured? How many history, anthropology, and science examinations have you written? Life seems to be one test after another. There are even driver's exams, psychological profiles, Civil Service exams. You can't get into the army without being tested. Some of the most popular television programs have to do with how much we know about this or that, trivia they call it. We have even developed measuring sticks of how smart, or dumb, we are in comparison to everyone else: an IQ test we call it. We are truly obsessed with our intellect.

Other animals don't escape our curiosity about intelligence, either. Animal behaviorists want to know how smart animals are. They test everything from planaria, squid, cockroaches, mice, and elephants. They put them through mazes, teach them encoded messages, and test them on color, word, and shape discrimination. Biologists even seek to discover if other animals have a "self-identity," if they can solve problems, have reasoning ability, or even if non-human animals have a moral sense. Perhaps we will find out that some mouse somewhere is supersmart, the all-time maze champion. Perhaps we will uncover the Einstein among dolphins, or maybe we will find a chimp that belongs in college (he or she would, of course, have to take the SAT, or ACT).

So, what's all this fuss about intelligence anyway? Well, the fact is...mammals are smart. Among all animals that have ever existed, mammals have evolved the greatest degree of intelligence. The contest isn't even close. While insects dominate because of size, reproductive prowess, and instinctive behavior, mammals dominate because of intelligence. Birds have concentrated their evolutionary gifts into flight. Their brains reflect it: large optic lobes for acute vision and a large cerebellum for muscular control and coordination. Mammals, on the other hand, have invested their evolutionary heritage into a large cerebrum...the reasoning center, the problem-solving center, the communication center, and the thinking center. *A huge cerebrum is the hallmark of mammals.* It, as much as hair, or mammary glands, is their distinction.

Non-human mammals have complex behaviors, emanating from their large cerebrum, that enhance survival. From birth they are **inquisitive**. They explore every inch, every nuance of their terrain. They, more than any other animal on earth can, learn. Much of their learning is Pavlovian or **conditioned**, that is when a normal response to one stimulus becomes associated with a new stimulus. But for higher mammals, especially primates, learning can be of the highest order, due to **insight**. There is even some convincing evidence that chimpanzees have some awareness of self.

However, the crowning achievement of intellect is the human. The human mammal survives, not because of its size, speed, strength, or natural weapons. We survive because we have been given the bonus of vast cerebral hemispheres... we survive because we are smart.

Topic 24

Extinction

Nature is often hidden; sometimes overcome; seldom extinguished.

-Francis Bacon

Looking Back...Looking Forward

Animals blanket the earth with an amazing tapestry of color, shape, and design. We can marvel at their simplicity and their complexity, at their strangeness and their beauty, their clumsiness and their inventiveness. The ways they've adapted to life in every crack or crevice is nothing short of astonishing. With the help of evolution's sculpturing hand, they have invaded every available ecological niche, resulting in about 10 million different kinds. If that's so, then there's little doubt that more animals exist now, and in greater variety, than at any time in the history of the world. From sponges to squirrels, the success of animals is incredible. We can take heart that in the midst of the modern world more animals are out there than ever before.

While that is true...there *are* more animals than ever...they are not on safe ground. In fact, I would be remiss if I didn't point out that animals are, in fact, engaged in a struggle of mammoth dimensions. Not since the time of the dinosaurs have they faced such a bleak future, and few are exempt. Frogs are in trouble, so are butterflies, snakes, whales, and yellow-headed blackbirds. Pandas are under assault, as are owls, corals, and sharks. The threatening cloud enshrouds snails, lobsters, and shrikes alike. Not many escape the entwining reach of extinction's

tentacles. A war is being waged and the very survival of animals, and perhaps our own, lies in the balance.

To truly understand what is now happening to animals we need first to know something of extinctions. For example, are they a common phenomenon? Are they a recent thing? What causes them? What can we expect in the future? Is there anything we can do?

First the history. Scientists estimate that over 50 billion kinds of organisms have inhabited the earth since that first spark of genesis some 4.5 billion years ago. Fifty *billion* different species have lived and thrived here on our little planet. Well, if that ancestral figure is anywhere near accurate, and if our estimate of 10 million living animals is also correct, then that means that well over 99% of the species that once lived and prospered are no more. That's a sobering thought. The truth is, *extinction is normal*; each species appears, thrives, and then disappears. On average, a species has a life span of about four million years, then it's gone. Extinction, it seems, is the fate of every species. It is a process that has occurred on a slow steady pace over the eons. It has even occurred in massive episodes wiping out virtually all existing life-forms in but a blink of time's tired eye. In fact, we believe that at least five global extinctions have occurred, the last about 65 million years ago, the so-called K-T event, which eliminated approximately 70% of living forms, including the dinosaurs. *Moreover, we are convinced that the sixth such episode is now under way.*

But Mother Nature is a persistent sculptor... always creating with a slow, inexorable hand. And the tool she uses in her creative effort is genetic diversity. If the fossil record is being read correctly, then great extinctions have offered great opportunities. After each major episode, when the slate was just about wiped clean, new life-forms evolved and in greater variety than before. If nothing else, natural selection is expedient, taking advantage of any new void. Extinction then is nature's way of turning corners on a map-less road—a way to start off in a new direction, leaving behind many of the past efforts, even successes. When one species exits the scene, genetic variety makes it possible for another to take its place. The process is wondrous in its simplicity. As long as there are survivors, there is raw material for new life-forms. Replacements are just waiting on the sidelines, waiting in the genetic fabric of the DNA molecule; waiting in mutations yet to occur.

So, if extinction is the rule in nature's celestial game, why should we be so concerned? After all, it's a natural process and nature will take care of herself...she will create new and perhaps even more wonderful species. Why be concerned? Why?...for two reasons. First, natural selection is slow, real slow. Replacement takes millions of years, far beyond our limited ability to just wait it out. Second, and more foreboding, is that as wondrous as natural selection is, it's just as heartless. It cares not for what disappears, nor for what reappears. Perhaps in our zeal to have dominion over the earth, we will erase the very diversity upon which we depend. Perhaps in the chaos, we will become one of the losers, to be replaced by some unknown, maybe by some unthinking, uncaring new species. That's why we should watch very carefully what's happening to our earth's biodiversity.

Causes:

How does extinction occur?

There are really only two possible causes of extinction: that which occurs as the result of human actions and that which has other causes, unrelated to us. We will call one *natural* and the other *man-induced.* Keep in mind, however, that we are certainly part of the natural order and thus not really separate from nature's forces. But, for purpose of comparison we will use these two rather artificial, but real categories.

Natural Forces

If you think about it, virtually all of earth's acts of extinctions have nothing to do with humans. At least for the first five major episodes, we were not the culprit; they all occurred long before we came on the scene. We can't be blamed for the disappearance of trilobites. Nor should we have any guilt over the demise of dinosaurs. We didn't have a hand in the passing of archaeopteryx. We weren't there, we weren't involved...don't blame us.

So, what were the causes, then? Something was surely responsible for the loss of all those billions of species. What was it? Well, there are four or five possibilities. One is **predation**. A predator may be so efficient that it devours every member of a given prey species. We can

Figure 24.1 The likelihood that predation of one species on another, such as in this lioness feeding on the zebra, can cause large-scale extinctions is questionable. A much more likely culprit is competition. Whether it's interspecific (as between the hyenas and vultures above) or intraspecific (between the individual hyenas or between the individual vultures), competition is a potent force in the race for survival.

Figure 24.2 Cosmic forces, such as volcanoes, meteors, and temperature changes, can cause havoc on entire populations of living organisms.

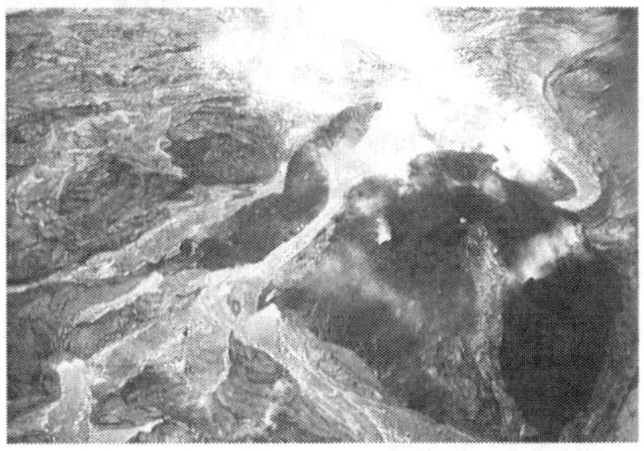

surmise that this surely happened here and there, but we can be equally certain that it didn't happen on a large scale. The predator-prey relationship is such that the two co-evolve, thereby reducing the likelihood that prey extinction *en masse* would result. **Parasites**, too, may have caused local extinctions, but not likely in great numbers.

A more likely possibility is **disease**. Diseases can be so invasive, so intrusive, so virulent that an entire species could succumb. There's little doubt that epidemic diseases have taken their toll throughout the ages. The extent, however, is again unknown, although it's doubtful that disease, by itself, has spawned extinctions on the grand scale seen in the fossil record.

A better candidate for large-scale extinctions is **competition**: both interspecific and intraspecific types. Competition for limited resources can be a compelling force in the fight for survival. Those species or those individuals better equipped in the race for resources will survive, often at the expense of others. This is without questions one of the prime forces in what's called **background extinctions**—that is, in the slow, gradual extinctions that occur constantly between the major episodes.

If predation, parasites, disease, and competition are only responsible for background extinctions, what then has been the cause of mass extinctions of the sort suffered by the dinosaurs? Our best guesswork to date is that massive disappearance of life-forms was caused by **cosmic forces**: volcanic activity, temperature shifts, asteroids, and meteors. Each of these alone can have grave and widespread effects on living organisms. If two or more happened simultaneously, the aftermath would be catastrophic. A volcano on an island can virtually extinguish all life within a few hours or days. A large one or a meteor the size of a few football fields, however, will have global consequences. The impact would most likely produce a shock wave and a poisonous cloud that will suffocate most land animals. The resulting temperature change would wipe out whole continents of animals. Sea level may drop, bodies of water may completely vaporize, photosynthesis may cease. Surely, the result would be cataclysmic.

Human Causes

Then along came man, a newcomer on evolution's playground. The best estimates place our emergence at about 500,000 years ago. We burst on the scene as a new species, *Homo sapiens,* a species destined to dominate earth as none had before, or since. Humans entered the extinction equation and changed the rules. We, with our huge brains, our sense of self-awareness, our unique concept of morality, our immense desire to consume, have altered the very face of our planet.

Just like a massive asteroid, we have collided with earth. And we are only now beginning to comprehend the weight of that impact. Like a horrendous meteor, we have issued forth our poisonous cloud of pesticides, hydrocarbons, and acid rain. We have raised and lowered water tables, sent shock waves of atomic explosions, denuded the prairie, slashed the forests, and we have even, it seems, altered climate through global warming. What will be the price other animals must pay for *our* time here?

Based on all the evidence available to them, scientists estimate that, on average, 10 species **should** suffer extinction each year. At the end of the 20th century, that number reached about 10 per day! *That's nearly 400 times what we would expect.* A staggering change had taken place and has catapulted us head over heels into the sixth major episode of worldwide extinction. But are we really to blame? Consider the following:

Figure 24.3 Humans are having a devastating impact on other species. Whether it's through burning rain forests to clear land for agriculture (**A**) or through polluting waterways (**B**), the effect of human activities is drastic. The escape of domestic animals (goats, horses, cats, dogs, pigs, even frogs), which become feral, can be equally devastating by quickly degrading otherwise stable habitats (**C**-**E**).

(1) **Habitat depletion.** Since our appearance on earth, 50% of wetlands and over 90% of forests are gone. Millions of acres of grasslands and prairies have vanished; rain forests, those genetic safety nets where most of the planet's diversity resides, are disappearing at the rate of hundreds of acres each day.

(2) Our air, waterways, and soils are **polluted** with deadly toxins. Tankers and oil platforms break open and oil covers sea stars and sea birds, agricultural runoff sours our streams, garbage heaps choke our groundwater, dams flood pristine habitats, and acid rain showers ponds and lakes with their deadly mists.

(3) In our effort to control pests, to enhance domestic livestock, or even just to diversify habitats, we have intentionally (and sometimes accidentally) introduced new species into areas where they often out-compete natural flora and fauna. As a result, introductions of "exotics" have sometimes had disastrous consequences.

(4) In the name of sport, food, clothing, or other "comforts," we have **exploited** animals without regard. Passenger pigeons vanished because of our taste for pickled pigeon, the dodo was clubbed out of existence, and alligators nearly paid the ultimate price for shoes or handbags. Whales have suffered, fish populations have declined, otters have given their fur for fashion, egrets for feathered head wear, elephants sacrificed tusks for piano keys, and rhinos donated their horns for sexual pleasure (ground rhinoceros horn is thought in some cultures to be an aphrodisiac). The list is a long one; so many other species are paying the price of our pleasures.

(5) Finally, there's the issue of **human population size.** Our species is near the seven billion count. That's a lot of people wandering about our planet, each one hungry, thirsty, and in need of a little space. And it's about to get worse. Experts claim that our numbers will double every 30 years or so. With food, drinkable water, and space being scarce right now, what will the future hold for us and for those other species sharing this planet? There's bound to be a squeeze. Which species will suffer most when the squeeze is applied? Your guess is as good as mine...but I'm afraid we've already seen a glimpse into that future as the number of animals nearing extinction grows dramatically day by day.

Figure 24.4 From feathers to furs, animals continue to be exploited for our comfort. As our numbers continue to expand, what price will future animals pay?

So What?

Let's face it, animals will suffer extinction no matter what we do. So what's the big deal? Why should we care? What's a species worth anyway?

Let me suggest a few ways to address such questions. First of all I, for one, don't want to live in a world without giraffes, pandas, alligators, or wild roses. I would rather not see owls, butterflies, koalas, and sea anemones vanish. I would much rather share this planet with caecilians, cougars, and eagles than just with cockroaches, rats, and starlings (although I guess they have their place, as well). The truth is, I relish nature's diversity, the more the better. And I would guess that I do not stand alone; you might very well share my preference for variety.

But there's another reason to care, and this one affects us all. Do you realize that over 50% of current **medicines** are derived from living organisms? Aspirin comes from willow, penicillin from a fungus, a northwestern tree has given us an ovarian cancer drug, and clawed frogs may very well ooze the next class of antibiotics from their slippery skin. After all, plants and animals have developed protective measures against disease over millions of years—measures that we are now just beginning to tap. But if a large proportion of wild species vanish, they may carry with them the secrets to solve today's diseases and those yet to surface.

Another way to answer the "so what?" question is the **cave-in principle**. I suspect that if certain key members of an ecological community are lost, they may just bring down the entire pyramid with them. There may be a domino effect when a critical species disappears, leaving others unable to adjust; if they vanish, the whole ecosystem may collapse.

And there are some other things to consider when you ponder the value of a wild animal species:

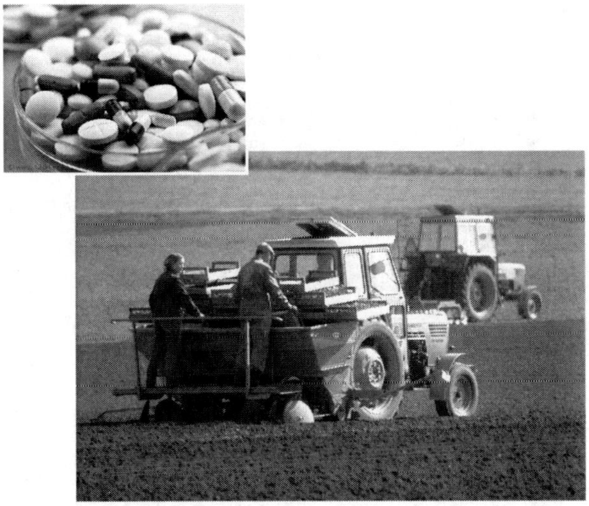

Figure 24.5 From medicines to food, plants and animals have contributed immeasurably to human welfare.

...other benefits from animals

√ **genetic diversity** - Each organism's DNA is unique and once lost cannot be recovered... *extinction is,* indeed, *forever.*

√ **environmental monitors** - Many species can warn of environmental decline (e.g., eagles and other large raptors warned us of the biomagnification effect of DDT; the death of pine trees signal acid rain, which can stagnate an otherwise productive lake).

√ **agricultural benefits** - Plants and animals are our prime source of food. As demand for new and better food grows, we will need to look to wild animals as a potential new source. But once they pass, we have lost that potential.

√ **educational** -We have an inherent tendency to seek information about the natural world. As animals disappear, the intricate ecological web is diminished and lost to our understanding.

√ **recreation** - Eco-tourism, bird-watching, hiking, camping, hunting, fishing, nature photography, and insect and shell collecting are all pleasures afforded by our natural world.

√ **moral imperative** - Many people believe that preventing extinction is "the right thing" to do. We are not alone here on earth and it is only ethically right to protect those species with whom we share this planet.

√ **survival** - The more we understand about plants and animals and their survival, the more likely it is that we can prevent or delay their extinction, and by extension, perhaps our own.

Solutions:

What can be done?

In 1973 the United States passed legislation to curb the alarming increase in animal and plant extinctions. Called the **Endangered Species Act**, it is a powerful tool to both identify and protect wildlife at risk. Since its enactment, thousands of species have been studied and hundreds have been "listed" as either threatened or endangered. It's difficult to gauge the success of such legislation; however, most scientists believe it to be an essential part of the war against needless extinctions. In fact the American bald eagle and the American alligator, two populations in trouble 30 years ago, have experienced substantial improvement to the extent that both species have recently been "delisted."

There's more...
...additional solutions include:

(1) reducing the size, or at least the rate of growth, of the human population, perhaps the greatest current threat to animal survival;

(2) setting aside large tracts of protected land as wildlife sanctuaries;

(3) reducing animal exploitation, including the control of animal harvesting for sport or commercial profit;

(4) controlling the use of biocides by intensifying efforts to use integrated pest management in our battle against agricultural and other pests, and

(5) being ever mindful of the possible damage done by the introduction of exotics into new environments.